WISDOM OF THE BEES

Wisdom of the Bees

Principles for Biodynamic Beekeeping

Erik Berrevoets

STEINERBOOKS • GREAT BARRINGTON, MA • 2009

STEINERBOOKS
An imprint of Anthroposophic Press, Inc.
610 Main Street, Great Barrington, MA 01230
www.steinerbooks.org

ILLUSTRATION BY THE AUTHOR EXCEPT WHERE INDICATED OTHERWISE.
BOOK & COVER DESIGN: WILLIAM JENS JENSEN
COVER IMAGES: APIS FLOREA NEST COURTESY OF SEAN HOYLAND (WIKIMEDIA COMMONS)
SUNSET IMAGE © BY JOSHUA HAVIV (SHUTTERSTOCK.COM)

LIBRARY OF CONGRESS CATALOGING-IN-PUBLICATION DATA

Berrevoets, Erik, 1960–
    Wisdom of the bees : principles for biodynamic beekeeping /
Erik Berrevoets.
        p. cm.
    Includes bibliographical references.
    ISBN 978-0-88010-709-9
    1. Bee culture. 2. Organic farming. I. Title.
    SF523.B438 2009
    638'.1—dc22

                                        2009026867

CONTENTS

I

Introduction: A Spiritual Approach to Beekeeping

"The more you investigate these creatures and the manner in which they live, the more you will come to the conclusion that there is a great wisdom in how they work and what they accomplish."

—Rudolf Steiner (120)[1]

Today, more than eighty years after Rudolf Steiner presented his lectures on bees, we are confronted with the decline of the honeybee. This alone justifies revisiting the lectures and reexamining his observations and insights for possible suggestions to rectify today's situation.

While the benefits of Steiner's research into agriculture and education are increasingly recognized, as evidenced by the interest in schools dedicated to Waldorf education and biodynamic gardening and farming practices, his research into the nature of bees has had limited impact on beekeeping practices and on our general understanding of nature. This book examines Steiner's insights and research into the nature of bees and their implications for beekeeping.

---

1. Rudolf Steiner, *Bees* (Anthroposophic Press, 1998), 8 lectures, Dornach, Feb.–Dec. 1923. All references to this text appear throughout the present book as page numbers only.

My personal involvement in beekeeping arose very much in response to the sad state of the honeybee around the world.[2] Despite my long-term fascination with honeybees, conventional beekeeping practices depicted in mainstream beekeeping literature suggest that the only way to manage bees is to dress in a fashion that, to the uninitiated, looks as if one were cleaning up a toxic waste site or disarming a nuclear power facility. Such a dress code, along with the use of industrial-type blowers to remove bees to facilitate taking the honey and utilizing carbon monoxide to anesthetize queen bees to facilitate artificial insemination, prevented me from getting involved. However, my fascination continued, and I decided to keep bees and explore alternative management practices. One of my first observations—that the bees were perfectly happy to let me observe them up close, while any attempt to open the hive and manipulate the frames was met with aggression—led me to conclude that the bees much prefer to be left undisturbed.

A search for alternative beekeeping practices among biodynamic and organic literature led to some very worthwhile books by beekeepers such as Ross Conrad (2007), Gunther Hauk (2008), and Michael Weiler (2006). Their books contain a wealth of information on the necessity of adopting a more ethical attitude for dealing with bees—natural approaches to addressing the effects of pests and diseases, sympathetic

---

2. The honeybee is said to have originated in Southeast Asia, from where it moved north as far as Northern Europe. Over thousands of years, honeybees developed numerous subspecies that adapted to local conditions. The issues affecting the honeybee are important for humanity, because the honeybee made possible the spread and dominance of European agriculture around the world during the colonization of the Americas, Asia, and Oceania. The destruction of many indigenous forms of agriculture and knowledge of indigenous food plants, together with the homogenization of agriculture in general under the influence of Western science means that today humanity has become increasingly dependent on a relatively small number of animals and plants for its food production. A great number of these rely on the European honeybee (*Apis mellifera*) for pollination. Effectively, humanity has become largely dependent on the health of a single insect species for its food security.

approaches to reducing our reliance on chemicals, and a need to avoid practices generally associated with agribusiness. This literature, unfortunately, contains little on managing bees without regularly manipulating the hives.

One publication stood out, however. It contains a series of lectures by Rudolf Steiner on the nature of bees, given in 1923. The lectures did not prove to be an easy read, even for one with an understanding of biodynamic agricultural principles. However, after several readings, the book revealed an entirely different way of viewing bees and nature. At first glance, the lectures do not seem to offer much practical information of immediate use. Steiner's better known agricultural lectures of 1924 on biodynamic methods have a more practical focus. Nevertheless, the lectures on bees contain a deep spiritual understanding of the wisdom that underpins all of nature. They are a treasure trove of information on the nature of bees, the impact of economics and politics on beekeeping, the importance of honey for people, and the importance of bees for plants, all of which can be used as a basis for developing alternative beekeeping practices.

One of the key findings Steiner presented in those lectures is the need to understand a whole bee colony as an individual organism rather than as a collection of independent animals. As with any living organism, a bee colony objects to disturbance and exposure of its internal processes; this is clearly indicated by the bees' aggression toward those who attempt to do so. This finding resonated strongly with me and resolved my initial desire to find a way to keep bees without agitating the colony and in a way that respects the integrity of a bee colony as an organism.

Steiner's lectures also offer insights into the causes of the pests and diseases that affect bees, while pointing to possible alternatives to conventional approaches that seem unable to solve problems and often lead to others.

Steiner's research shows not only the interconnectedness that exists within a bee colony, but also the interconnectedness between insects and plants. He found that this interrelationship provides mutual benefits for both the insects and the plants. This observation is particularly important because, in contrast to the increasing attention given to the dying oceans and river systems and the decline of animals and plant species, little attention is given to air-based ecosystems and insects. Bees are one of only a small number of domesticated insect species, which also include silkworms (*Bombyx mori*) and lac beetles (*Laccifer lacca*), which produce the material for making shellac. Perhaps this is the reason human beings are generally not very concerned about the state of insects. In fact, most people would regard the decline of insects as beneficial rather than a matter of concern. Nevertheless, if the honeybee's decline is indicative of the decline of other insect species, this could have as yet unknown consequences for the world we live in.

Rudolf Steiner's Lectures on Bees

In 1923, Steiner presented a series of lectures on bees, wasps, and ants to the construction workers at the Goetheanum in Dornach, Switzerland. They followed a presentation on practical beekeeping given by a man identified as Mr. Müller, a beekeeper at the forefront of beekeeping practices such as artificial breeding. As such, Steiner did not feel the need to discuss practical aspects of beekeeping; instead, his lectures are more theoretical and discuss his spiritual research into the nature of bees, his views on the beekeeping practices presented by Müller, and his responses to questions from the beekeepers in the audience. As a result of Steiner's answers to these questions, many issues are revisited and expanded upon throughout the lectures instead of being grouped together. This sometimes makes it difficult to understand immediately Steiner's findings on certain matters.

Steiner's lectures on bees are multilayered. On one level, they present information based on his research on bees and other insects.

On another level, the lectures provide examples of his methodology for studying the spirit and wisdom expressed in nature. As such, the lectures have a number of objectives, the first of which is to encourage his audience to develop a deeper understanding of bees and nature by cultivating one's awareness of how the nonmaterial reality, or spiritual dimension, affects the sense-perceptible world. Another objective is to show the importance of bees and other insects for humanity because of their role in revitalizing plant life, pollinating food crops, and producing honey.

In the prelude to his lectures (February 3, 1923), Steiner discussed how spiritual research can reveal the unique nature of bees as the outcome of certain effects of the cosmos on the bee colony, in particular the influence of Venus. He also showed how the bee colony's unique means of reproduction and its social organization are reflected in the special relationship bees have with flowers. The spirit element that bees derive from flowers when they collect the nectar and pollen on which they feed creates the bee colony's unique characteristics. This spiritual element is also present in the honey.

In the first lecture (November 26), Steiner further discussed cosmic influences on the nature of bees. He showed how the various types of bees that make up a colony are a result of mediating influences from the Sun and other cosmic forces through the hexagonal shape of the comb cell, as well as variations in the duration of their exposure to such influences as the bees develop from egg to full maturity. He also showed how such influences affect other aspects of a bee colony, such as swarming. Steiner focused on the similarities between the bee colony and the human body to better understand the functioning of a bee colony. After explaining the internal characteristics of a bee colony, he examined the relationship between bees and plant life to show how there is evidence in nature of a kind of wisdom. Understanding this wisdom is the first step in assessing the possible effects of conventional beekeeping practices such as artificial bee breeding on the long-term well-being of a bee colony. Steiner stated that, if such practices ignore the bee colony's natural processes, the long-term impact will be negative.

In the second lecture (November 28), Steiner responded to an article on bees' ability to see colors. He criticized the conclusions that conventional scientists draw from their research on bees and nature. His critique is centered on researchers' assumptions that the effects they observed have the same significance for animals as they have for humans. Steiner indicated that the chemical processes perceived through smell and taste assume greater importance for animals than they do for humans. His other criticism of conventional science is that, by controlling scientific journals, scientists determine what is to be considered true and hinder access to alternative approaches. Steiner concluded that, as a result of the manner in which conventional science conducts research, it has very little of practical value to offer beekeepers. During this lecture Steiner also commented on the importance of bee nutrition for their well-being and the effects of artificially feeding bees.

In his third lecture (December 1), in response to an article discussing the benefits of honey cures for children, Steiner pointed out the shortcomings of conventional science with respect to nutritional research. He argued that nutritional research needs to adopt a holistic perspective and that, instead of explaining the benefits of specific foods by breaking them down into their chemical components, it is important to explain the benefits derived from the characteristics of the spiritual elements behind their creation. He explained the meaning of this by discussing the benefits of honey for people, referring to the unique fact that it originates as nectar produced by plants, is then gathered and transformed by bees into honey, after which it is stored in six-sided honeycomb cells. He also responded to questions from his audience on the connection between bees and people and on the planetary influences on honey production.

Steiner elaborated on the connection that exists between a bee colony and a beekeeper in his fourth lecture (December 5). He focused on the ability of bees to recognize their beekeeper. He explained that, contrary to our common understanding, a bee colony should be regarded as an entity, or organism, with its own memory, and not as a collection of individual bees. Steiner explained that beekeeping practices are developed in response to the political and economic conditions of a particular period. He discussed conditions associated with the general commercialization of society, conditions characterized by a short-term focus on profit and quantity over quality. This led to beekeeping practices such as artificial breeding, which have short-term economic benefits while ignoring the long-term effects of altering the natural processes of bee colonies. Steiner also answered questions on why some people cannot tolerate honey and on the influence of zodiacal signs on honey production.

The fifth lecture on bees (December 10) focused predominantly on Steiner's answers to issues raised by the audience, including the connection between the prevalence of bee diseases and artificial bee breeding, feeding, and the use of chemical fertilizers. Steiner again pointed out how interfering with a bee colony's natural processes has negative effects. He also expanded on the importance of spiritual science for understanding the natural world, discussing the relationships among various types of wasps and trees and the origins of honey. He explained how, in the past, people used their understanding of nature to exploit the relationship between fig trees and wasps to create figs with greater nectar content. He commented that such insights were instinctive in ancient times, but that such understanding can now be gained through spiritual research.

In his sixth lecture (December 12), Steiner revisited the relationship between bees and flowers and the importance of honey for people. He also responded to questions on the effects of bee venom, as both a cure and a poison. He explained the

importance of considering a person's individual characteristics
and medical condition when deciding on a course of treatment,
warning against general statements about the benefits or otherwise of the substances used in treatments. Steiner also discussed
how one can change a worker bee into a queen bee by manipulating its nutrition, the shape of its cell, and the duration of its
development. Both Steiner and Müller warned, however, that
such a bee should not be considered a healthy queen bee, but
rather evidence of an unhealthy colony. Steiner discussed the
wisdom behind natural processes, explaining the differences
between bees and insects with a similar social organization
(such as ants and wasps) by pointing to the differences in nest
structure and nest materials.

Lecture seven (December 15) goes further into the differences
among bees, wasps, and ants. Steiner said that to call the complex nest building and reproductive behavior of insects instinctive does not explain anything, but merely gives those activities
labels. Instead, he insisted that such behavior is evidence of the
wisdom and spiritual elements that underlie natural processes.
He pointed out that there is a unique relationship between
insects and plants, a beneficial relationship for both, which cannot be fully understood without spiritual science. Steiner also
discussed the need to focus on the historical development of
plants and animals to understand their present characteristics.
He concluded that civilization has become too much of a brain-
based culture at the expense of intuitive understanding.

In his final lecture on bees (December 22), Steiner discussed
the mutually beneficial relationship between insects and plants,
focusing this time on the physical aspects of that relationship, in
particular the role of oxalic acid, formic acid, and other insect
venoms. He discussed, too, how these play an important role in
the breakdown and revitalization of the earth, plants, insects,
and the human body. He also explained the significance of bee
venom in the bee colony's swarming process. Steiner concluded

this last lecture, which was held around Christmastime, by indicating the true symbolism of Christ in a moral sense, the re-enlivening of the earth.

Overview of the Book

In addition to this series of lectures on bees, Steiner also referred to bees in several other lectures, and some of that material has been included in this book. The lectures on bees assume some knowledge of beekeeping and Anthroposophy, and this information has been added as context to provide a better understanding of the material in Steiner's bee lectures.

At times, Steiner expressed concern about the use of his lectures as representative of his overall views on certain matters. He preferred to review and revise his lecture material prior to publication so that he could explain any issues that might be viewed as ambiguous or in need of clarification. Unfortunately, his death in 1925 prevented him from doing so with neither the lectures on bees nor those he presented in 1924 on agriculture.[3]

The multilayered nature of Steiner's lectures means that it is possible to lift out a range of different aspects for discussion and contemplation. The artist Joseph Beuys, for example, derived artistic inspiration from Steiner's discussion of the similarities between a bee colony and the human body.[4] Others have taken his findings on the intimate connection between the bee colony and its implication for the development of society as a theme for further development.[5] The focus of this book is very much on the findings of Steiner's spiritual research into the nature of bees and their implications for beekeeping,

3. *Agriculture Course: The Birth of the Biodynamic Method* (London: Rudolf Steiner Press, 2005), 8 lectures, Koberwitz, June 7–16, 1924.

4. See "From Queen Bee to Social Sculpture: The Artistic Alchemy of Joseph Beuys," an essay by David Adams, in *Bees* by Rudolf Steiner.

5. In March 2008 the West Australian Branch of the General Anthroposophical Society held an international event entitled "The Bee Master: A Michael Easter Festival," which focused on the social application and implications of aspects of Steiner's lectures on bees.

sometimes at the expense of discussing and presenting some of his more esoteric observations and comments.

Throughout his lectures Steiner revisited a number of themes, which have been grouped together and form much of the first few chapters of this book. Following this introductory chapter, the following three chapters discuss Steiner's findings on the nature of bees. These are followed by four chapters that discuss the effects of conventional beekeeping practices against the background of Steiner's research on bees and provide suggestions for beekeeping practices based on his understanding of bees.

Chapter 2 begins by discussing how social and economic conditions shape beekeeping practices. At the time of Steiner's lectures, beekeeping in Europe was beginning to change in response to the increasing commercialization of society. Those changes were based on a desire for increased returns through increased honey production. This was made possible by new beekeeping practices such as artificial bee breeding. The ability to increase honey production overshadowed any notion that pushing bees to achieve this could have long-term negative consequences.

Steiner pointed out that the failure to understand the impact of such practices is reinforced by the way conventional science studied nature, which provides only a partial understanding of nature. Steiner showed his audience that, by studying the spirit element that forms the material basis of nature, we can gain a more appropriate understanding of the nature of bees, against which one can assess the effects of beekeeping practices. He referred to his method as spiritual science. Steiner's approach to understanding nature is discussed in the second part of the chapter and provides a brief outline of what studying nature spiritually involves and how this differs from more conventional approaches.

Steiner's spiritual research on the nature of bees makes it very clear that practices such as artificial bee breeding will have long-term negative effects for the health of bee colonies by strongly interfering with the natural reproduction of the bee colony.

Chapter 3 reviews Steiner's spiritual research by discussing the influences of the planet Venus on the intricate relationships that evolved among the various types (or castes) of bees within the colony, and he explains how such influences are reflected in the social organization and reproduction of the bee colony. The chapter discusses, too, how the influence of the planet Venus on a bee colony also extends to the relationship between bees and plants, since bees derive all their nutrients from that part of the plant that is most strongly associated with Venus: its flowers. Through bees, the influences of Venus find their way from plants into honey, and by consuming honey those influences also benefit humans.

Steiner's findings show how the differences among the bees in a bee colony are the result of environmental factors and how, by manipulating these factors, eggs intended to develop into worker bees can be changed into queen bees. At the time, this was not regarded as having any implications for practical beekeeping, and even progressive beekeepers considered queen bees raised in this way to be inferior to those developed naturally by the colony. Since then, however, rearing queen bees from worker eggs has become common practice throughout the Western world.

Chapter 3 also discusses the similarities between a bee colony and the human body as further evidence of the interrelationship among the different bees in a colony and how the role of the queen bee, worker bee, and drones in a colony are similar to that of the protein, blood, and nerve cells in the human head.

The relationship among the individual bees within a bee colony has often been used as a model for human society. However, Steiner's research shows that individual bees are more similar to the cells of a human body than to individual people. He concluded that a bee colony should be understood as an organism rather than as a collection of individual animals.

Chapter 4 discusses Steiner's point on the importance of examining the intimate relationships in nature holistically, by focusing on cooperation and mutual benefits rather than on competition and survival of the fittest. This is explained in connection with bees and flowers and the importance and benefits of bees for humanity. Bees ensure the continuation of plant life on Earth, spiritually and physically, at a time when the Earth is losing vitality as part of its current phase of evolutionary development. Bees not only enable plants to reproduce through pollination, but they also enable plants to remain vital by depositing bee venom during pollination, as bee venom has therapeutic properties that prevent plants from dying.

In addition, bees benefit people by producing honey. Honey enables people to benefit from the cosmic forces captured by the plants and those imparted by the bee when nectar is transformed into honey, as well as the forces that come from the hexagonally shaped comb cells in which honey is stored. By studying the similarities between a bee colony and the human body, Steiner explained that the same forces that enable bees to make six-sided wax cells in the honeycomb can also be used by human beings to construct their own bodies.

Chapter 5 discusses Steiner's research and that of others with a similar sensitivity to the nature of bee colonies. This research shows the wisdom that underpins the relationships between bee colonies and their external environment in the selection of their nest sites, and that the comb is constructed in a particular manner to create the optimum internal environment for raising brood. By comparing the bee colony's natural processes with conventional beekeeping practices, it quickly becomes clear that those methods largely ignore the wisdom of the bees.

The chapter shows, too, how conventionally framed hives dramatically interfere with the bee colony's natural processes of comb-building and the creation of a stable internal hive environment to raise its brood. The use of framed hives and associated management practices disrupt the bee colony's internal functioning, requiring additional effort to maintain a stable internal environment, which is essential for the proper development of its brood. These conventional framed hives, therefore, do not appear to be in harmony with beekeeping practices that work with the bee colony's natural processes.

The chapter concludes by discussing features for alternative hive designs that are more respectful of the bees' nature, such as top-bar hives and the hive developed by L'Abbé Émile Warré (ca 1876–1951). It is acknowledged that developing a hive that respects the integrity of a bee colony and that meets the objectives of the beekeeper is not an easy task, whether technically or in keeping with legislated restrictions in certain jurisdictions. However, by incorporating these features in hive designs and associated management practices, the bee colony's natural processes are shown greater respect and the shortcomings associated with conventional framed hives are addressed.

Chapter 6 focuses on the effect of artificial queen bee breeding and rearing. At the time of Steiner's lectures, when such processes were far less advanced than they are today, he nevertheless regarded them as having a major negative impact on the long-term fertility of bee colonies.

The principles that underpin artificial bee breeding and rearing contrast sharply with the principles underpinning the study of nature from a spiritual perspective. Conventional practices show little regard for the wisdom behind the bee colony's natural reproductive processes. Conventional bee breeding focuses predominantly on manipulating the queen bees and their genetic makeup rather than on working with the holistic relationships that already exist between the queen bee and the bee colony and between the bee colony and its local environment.

Rudolf Steiner's research shows that an organism's characteristics are the outcome of its response to the external environment, and that the better the match between the external environment and the organism, the better it is able to show its potential. Beekeepers influence the bee colony's environment through the choice of beekeeping practices, and in doing so they are able to bring out the best qualities of their colonies in that particular environment. The chapter also shows that a spiritual approach to bee breeding recognizes that economic, social, and environmental conditions change, and that bee breeding strategies should enable bees as a species to adjust effectively to such changes by maintaining diversity.

Chapter 7 discusses bee health, pests, and diseases from a spiritual-scientific perspective. Steiner found the negative effects of pests and diseases to be the result of an imbalance in the bee colony. It is important to recognize that bee colonies have developed natural processes to deal with pests and diseases by creating and maintaining a stable internal nest environment and by using their venom against intruders.

To restore balance in a bee colony, Steiner argued it is essential to restore its natural processes and to compensate for the negative effects of conventional management practices. This approach differs from conventional science, which does not generally link the occurrence of pests and diseases to interference with the bee colony's natural processes. The inability of conventional science to explain and treat the most recent bee colony ailments, such as colony collapse disorder (CCD), points to the need to explore other avenues to minimize their effects.

Steiner paid particular attention to the role of nutrition in restoring balance and suggested a number of strategies to restore the health of a bee colony, including increased access to beneficial plants and enabling bees to forage in locations where biodynamic preparations are used.

In light of the severity of the current issues affecting honeybee colonies, it may be necessary to employ interim measures to address the immediate effects of pests and diseases; however, it is important to realize that these alone do not provide a cure and do not address the causes of disease or pest outbreaks. One needs to understand the well-being of bee colonies holistically and to assess the impact of any intervention on all relationships within the bee colony, as well as the colony's relationships with the wider environment.

The final chapter discusses the role of the beekeeper in the context of bees' importance for plants and human beings, both directly though the production of honey and indirectly through the revitalization of plant life and the pollination of plant foods. The objective of beekeeping is to ensure that bees are able to continue this important role for the long term. The beekeeper's responsibility is to support bee colonies in this objective by developing a proper understanding of bees both physically and spiritually, enabling a full assessment of the effects of beekeeping practices. Spiritual research emphasizes the importance of taking into consideration the unique characteristics of the natural, social, and economic conditions of the environment in this process.

My hope is that this book leads to future discussions of the application of Steiner's spiritual-scientific research on the nature of bees for practical beekeeping. In the words of Steiner: "Every human being should show the greatest interest in this subject because, much more than you can imagine, our lives depend upon beekeeping" (Steiner 1923, 5).

2

SPIRITUAL SCIENCE AND BEEKEEPING

"You'll be able to gain insight and reach a conclusion only by applying the powers of mind, intellect, soul, heart, spirit, and imagination. This is what 'looking at something spiritually' really means."

—RUDOLF STEINER (92)

Rudolf Steiner's lectures give us a rare insight into the changes affecting beekeeping during the early part of the twentieth century. In particular, the discussion between the two beekeepers, Mr. Müller and Mr. Erbsmehl, shows how the commercialization of beekeeping and the adoption of new practices to increase honey production were regarded with apprehension by some beekeepers and as positive by others.

At the beginning of Steiner's lecture of December 5, 1923, Erbsmehl was concerned about the increased commercialization of beekeeping, saying, "Beekeepers today are concerned primarily with the profits gained from beekeeping. Every effort is directed toward this material goal" (65). To support his claim, he cited the *Swiss Apiculture Newspaper* (October 1923), which mentioned that honey had become a luxury commodity and should attract a good price. He observed that these changes had occurred not only in relation to beekeeping, but also in society more generally: "Here in the heart of Europe, the goal is to get as much out of the honey as possible. An employer's primary interest is to see how much labor can be gotten out of the employees. The same applies to bees" (65). He pointed out that, in other parts of Europe, honey was not treated as just a commodity.

Erbsmehl's concerns were not shared by Müller, who considered the commercialization of beekeeping both inevitable and justified:

> It can be only a beekeeper with a small number of hives who doesn't sell any honey at all. Mr. Erbsmehl simply doesn't know yet how much is presently involved in a beekeeping operation and that a certain amount of book-keeping is necessary. If you don't plan financially for some profit in this venture, just as you would in any other enterprise, you'd simply have to give up beekeeping.

Müller explained that commercial beekeeping is possible only by adopting new beekeeping practices and that without these "the large quantities of honey available commercially would not be there if you didn't obtain honey that was produced by artificial breeding" (65–66). That is to say, Müller disagreed with an earlier comment by Steiner—that artificial bee breeding would lead to serious problems in the long term, despite the fact that such practices had been introduced only recently and had led to increases in honey production (66).

Steiner pointed out that the objectives of beekeeping are determined by the economic conditions of a particular period and described how, when he was growing up, beekeeping was part of the general farm activities and that little attention was paid to the time involved or the cost of managing bees. Similarly, people did not regard honey as a commodity, and honey was just as often given away as sold. Commercial beekeeping had come about relatively recently, he continued, with the introduction of wage labor and hourly rates of pay, which led people to attach a monetary value to their time and labor. As a result, these days beekeepers generally consider the financial cost of beekeeping and its impact on their income when it requires time spent away from paid employment. Such considerations are, of course, less important when bees are kept as a sideline or hobby.

Steiner also discussed how, under these economic conditions, beekeepers want to increases honey production, since they are paid for volume not quality. As Müller explained, to receive adequate monetary returns for their labor, beekeepers must sell a certain amount of honey. The greater the level of returns a beekeeper desires, the greater the amount of honey a beekeeper must sell. One way to produce more honey is to increase the amount of honey a beekeeper can extract from each hive. Consequently, beekeeping practices such as artificial bee breeding and feeding, which increase the amount of honey per hive, will be regarded favorably. However, Steiner explained, increases in production often occur at the expense of quality, and if healthier economic conditions were to prevail, then the price of honey would be determined by its qualities, and beekeepers would obtain a better price for their produce.

Commercialization of beekeeping also led to a short-term focus, with the financial returns of an enterprise typically calculated for a year or less. This meant that the long-term effects of beekeeping practices could be ignored in favor of short-term economic returns. This attitude was reinforced by government authorities and through beekeeping journals such as the *Swiss Apiculture Newspaper*. Conventional scientific research, too, supported the adoption of beekeeping practices such as artificial bee breeding and feeding, which produce short-term economic gains. As a result, such practices continued at a great pace throughout the twentieth century and into the twenty-first (also see Baker 1948). Beekeepers such as Erbsmehl who questioned these practices were regarded as hanging onto tradition and standing in the way of progress.

Such developments were not unique to beekeeping; in Europe and elsewhere, traditional farming methods were also considered to be hindering progress. Against this background, it is easy to see that beekeepers like Müller did not understand what Steiner meant when he said that, "a century from now, all breeding of bees would cease if only artificially produced bees were used" (178).

Steiner acknowledged that it might be difficult to accept his criticism of certain beekeeping practices without understanding how he arrived at his conclusions. He agreed, too, with the impossibility of determining the accuracy of his conclusions without the relevant data. He explained that his comments are based on his understanding of the bees' nature through spiritual research, which is qualitatively different from conventional scientific research. He explained that, when certain beekeeping practices are introduced they alter the natural processes of a bee colony, and that any negative effects of this may not show up immediately, as living beings are, to some degree, able to adapt to external interference, but that this takes place slowly over a number of generations: "Something might be quite healthy at one time, but it will still show aftereffects later" (74).

Rudolf Steiner also acknowledged "it is impossible to object in any way to the artificial methods applied in beekeeping. This is because we live in social conditions that do not allow anything else to be done" (75). However in relation to practices that increase honey production, he warns,

> Nevertheless, it is important to gain this insight—that it is one matter if you let nature take its course and only help to steer it in the right direction when necessary, but it is entirely another matter if you apply artificial methods to speed things along." (75)

Steiner advocated research into the effects of modern bee-keeping practices to arrive at a full conclusion of their impact, suggesting, "Let's talk to each other again in one hundred years, Mr. Müller; then we'll see what kind of opinion you'll have at that point. This is something that can't be decided today [DECEMBER 5, 1923]" (75).

More than eighty years after Steiner's insights, the survival of the honeybee has reached a critical point in many areas of the world. Conventional science seems unable to explain the situation or offer solutions. By contrast, Steiner's research led him to conclude that certain beekeeping practices would lead to the negative results we observe today. Perhaps an understanding of his approach to studying nature and a reexamination of his findings on the nature of bees will provide a way forward and enable the honeybee to survive and for us to adopt an alternative way to live with nature.

Spiritual Science and the Study of Nature

Steiner presented his lectures on bees not only to share his findings, but also to educate his audience about the nature of spiritual research. He did not explicitly outline his research methods in those lectures; rather, he believed that, by presenting facts in a particular manner, his approach would become self-evident and more readily accepted. Nonetheless, his research is not easy to understand without some background. By complementing the information from his lectures on the nature of bees with material from his other works, it becomes possible to arrive at a basic understanding of his approach. The following discussion is intended to provide a brief background and summary of spiritual-scientific research as applied to the study of nature.

Spiritual science (Anthroposophy) developed in part as a response to Steiner's criticism of how nature was studied scientifically in his time. One of his key concerns was that such studies did not explicitly acknowledge the qualitative difference between living beings and inanimate objects. Steiner believed that conventional science had lost sight of the very essence of what makes a living being alive. The failure to acknowledge this qualitative difference led to the belief that nature could be studied in the same way one studies inanimate objects, which Steiner considered inappropriate.

He also questioned the assumption that a living being can be reduced to the sum of its components, and that it could be understood by studying its individual components in isolation from their interrelationships, an approach often referred to as reductionism. He further objected to assumptions that a living being can be studied in isolation from its environment and that laboratory results are considered applicable to the larger world. He felt that the relevance of laboratory research for understanding nature is limited, since it ignores influences from outside the laboratory on the living being and fails to consider a living being's ability to engage with its environment. Associated with this, Steiner objected to the assumption that a living being can be studied in isolation from its evolutionary development. Steiner was also critical of the preference of conventional science for considering only those characteristics that can be quantified and measured as relevant to an understanding of nature, and for ignoring those that cannot. As a result, Steiner felt that materialistic, natural science could provide only a partial understanding, as it is based on quantitative, one-dimensional laboratory studies that lack context and practical application.

While many of those criticisms hold true today with respect to much biological research, other fields such as environmental science have adopted a more integrated approach. Nevertheless, the study of what Steiner referred to as the spiritual element in nature remains largely ignored.

In response, Steiner set out to develop his scientific approach to nature based on the central premise that a non-material, spiritual wisdom is the basis of nature. The objective of Steiner's scientific approach was to understand how the spiritual dimension impacts on the material world. He saw the material dimensions of the world as expressions of a spiritual dimension. To understand how spiritual dimensions affect material reality requires an understanding of this spirit element. Steiner called this approach spiritual science—a systematic investigation of the spirit, the wisdom that shapes the material world. In the last of his lectures on bees, he said, "Things as they exist in nature are created with extraordinary skill, and you would have to admit that there is definitely evidence of intelligence and understanding" (144).

In the development of his spiritual-scientific approach, Steiner was greatly influenced by the ideas of the German philosopher and writer Johann Wolfgang von Goethe (1749–1832). Steiner outlined Goethe's ideas in detail in his books.

As with any skill, becoming proficient in Steiner's investigative methods requires training. The training begins by improving one's perceptive abilities, with a focus on *imagination, inspiration,* and *intuition,* terms that refer to three stages of perceptive faculties along the anthroposophic path of knowledge and are associated with the person's body, soul, and spirit. Training at this level involves mind exercises and meditation. One objective is to develop an appropriate conceptual perspective to construct the research project and to interpret its findings.

To understand nature through this method, Steiner suggested that a researcher should not think *about* the living being or observe it in a detached manner, but rather *merge with* it and imagine being it. This contrasts starkly with conventional science, which insists that the object under study must be observed objectively and considered external to the researcher. Conventional science makes no attempt to understand nature or a living being from within.

Intuition, a higher capacity that features large in Steiner's research, is generally not considered a valid method for scientific investigation, as it is assumed that the knowledge obtained this way cannot be scrutinized or subjected to further investigation. Steiner refers to such a view as "faith" and points out that his use of intuition as a research method differs from faith, since it is possible for others with similar training to arrive at the same understanding. Steiner argues that intuition, as a means of investigating the natural world, is therefore no less scientific than other, more conventional methods. Steiner promoted the importance of others verifying his investigations and findings, making it very clear that his findings and insights should not be accepted merely on the basis of his word or authority. Further details of this aspect of Steiner's scientific method can be found in his book, *How to Know Higher Worlds: A Modern Path of Initiation*.

Steiner's research methods consist of various approaches to perception, analysis, and interpretation of nature that address his critique of the conventional scientific approach of his day.

Holism

Steiner was most critical of the tendency of conventional science to reduce the study of nature and living beings to their individual components, often down to the level of chemical processes at the cell level, on the assumption that conclusions thus obtained could explain the complex interrelationships of the natural world. He pointed to the limited understanding that arises from such an approach, comparing it to explaining the operation of a compass by analyzing the properties of the pointer without reference to the Earth's magnetic forces. He argued that, rather than focusing narrowly on a living entity, it is important to study living beings as part of a whole system and to identify the spirit, or wisdom, that forms the connections among seemingly unrelated elements. Adjusting one's focus in this way thus includes influences on Earth that come from the planets and the wider universe, both spiritually and materially.

Development

In response to the way in which conventional science examines living beings in isolation from their environment and evolutionary development, Steiner stated the importance of being aware of the fact that the current expression of nature is the outcome of an evolutionary process. We need to understand the earlier conditions of existence and how spiritual forces gradually developed the material reality we observe today. Steiner illustrates what he means by saying that the conventional approach of science is like a visitor who comes from Mars to Earth and studies the shapes of human corpses, which would not exist if they had not been living beings first. However, it is not until the visitor can observe a living human being that it becomes clear why corpses are shaped as they are.

Rather than examining aspects of nature in isolation and at a certain point in time, Steiner explained that the key to understanding the natural world is to trace the unique shape of living beings today to their non-material inner principle. He argued that the material development of living beings flows from a spiritual concept or idea, an archetypical form he referred to as the *typus*. Perhaps one way to imagine the concept of *typus* is as spiritual DNA. The *typus* is not a sense-perceptible, actual living being or a static conceptual form; rather, it is more fluid and expressed in a potentially infinite number of forms through its interaction with the environment. In this evolutionary process, living beings take an active part in their own evolution.

Steiner was also critical of the view that evolution is based on competition among various species or living beings, and that the underlying principle of evolution is survival of the fittest. He perceived this as one-sided and, therefore, not the whole picture. He advocated instead the importance of identifying and understanding the role played by relationships based on cooperation. Steiner perceived nature as consisting of complex interactions that develop and are sustained according to mutually beneficial outcomes.

Similarities

Steiner also criticized conventional science for regarding as relevant only those qualities and characteristics that can be measured and quantified, stating that this, too, leads to a limited understanding of nature. He emphasized the importance of focusing on characteristics not easily quantified or measured, such as forms and shapes, as these play an important role in understanding natural processes. This aspect of Steiner's approach depends largely on a researcher's intuitive ability and skills.

The rationale behind the importance of identifying, analyzing, and interpreting similarities in nature, according to Steiner, is that these are evidence of the spiritual processes responsible for creating those shapes. Thus, by identifying similarities at a material level, one can gain a deeper understanding of the spiritual processes that created those shapes. By identifying the relationship between a particular shape and function in nature, Steiner found that the relationship also occurs where other, similar forms are present, as they express the same spiritual force or originate from the same *typus*.

In his lectures, Steiner discussed how natural silica is formed into solid six-sided quartz crystals, and how that form can also be identified in the honeycomb cells in which bees develop and store their honey. He concluded that such similarities arise from the same spiritual forces. The effect of these spiritual forces is that they create solidity. The similarity of shapes in quartz and honeycomb cells means that those forces will have the same effect on the bees and the honey, as these, too, developed in six-sided cells. As bees produce honey that is stored and developed in six-sided, it provides solidity and structure to the human body when consumed by people.

Thus, the application of spiritual science leads to an entirely different understanding of the importance of honey for human beings than would an analysis of, say, measuring and quantifying its chemical components.

Nature-Centered

Steiner highlights how conventional studies of nature that require researchers to be objective and separate from the object being investigated have led to a tendency to anthropomorphize nature. He means that researchers approach the natural world from a human-centered perspective. Steiner insisted that the researcher needs to merge with the object of study and develop awareness of how the world appears from the perspective of the animal or plant. The objective is to suspend, thereby, any interpretation of the world from a human perspective. This requires ever-present inquiry into whether the human view of situations or objects is in fact the only or most appropriate interpretation of phenomena.

Of particular importance for the researcher in adopting a nature-centered perspective is to be critical of approaches based on an assumption that plants and animals must experience the world and natural phenomena as do human beings. This requires awareness of the fact that human beings perceive the environment primarily through the sensory organs of sight, hearing, and touch, and that this may not be the case with animals. Animals may well perceive the environment in ways that are quite different from those of human perception, primarily using senses such as smell and taste. The fact that the sense organs differ between humans and animals means that even if they do perceive the same environment it may well have a different significance (35–37).

The example Steiner used in his lectures to illustrate his point is research into the ability of bees to see color as discussed in the *Swiss Apiculture Journal*. He showed how researchers assumed that bees experience and respond to the world as human beings do, with sight playing an important role in the life of a bee colony. He asserted that, by doing so, the researchers ignored the fact that, for bees, other senses such as smell and taste play a far more important role. Because such researchers remain unaware of their anthropomorphic assumptions, their findings have very little benefit for understanding bees and beekeeping.

The key differences between the investigative method developed for the study of nature by Steiner and the conventional scientific approach of his day is summarized in the table on the next page.

Discussion

During the time of Steiner's lectures, the beekeeping industry in Europe was at the beginning of radical changes in response to the increasing commercialization of society. These changes led to a focus on increasing financial returns by increasing honey production. The increase in honey production was made possible by adopting beekeeping practices such as artificial bee breeding. At the time, such developments were not welcomed universally by beekeepers, but support from government agencies and promotion in agricultural journals ensured that they would be widely adopted.

Steiner insisted that beekeepers should be aware of the economic and social dimensions of their time and recognize how they shape the development of beekeeping practices. He discussed how commercialization led to changes in the perception of time and labor, and that these influenced the way in which new beekeeping practices were assessed. The short-term focus on economic returns and the ability of certain beekeeping practices to achieve such great increases in productivity make it difficult to accept that such seemingly beneficial practices could eventually lead to negative consequences.

Steiner pointed out that the failure to understand the negative impact of such practices was reinforced by the way conventional science studies nature. He objected to the way those studies ignore the spirit element at the foundation of the material world, while viewing living beings increasingly as machines. He also objected to the way these studies lack a

CONVENTIONAL AND SPIRITUAL METHOD OF RESEARCH

CONVENTIONAL SCIENCE	SPIRITUAL SCIENCE
No significance is attached to the fact that a living being is alive.	Understanding that spirit, which gives a living being life, is central to understanding nature.
The aim is to be detached and objective.	The object of investigation is penetrated through intuitive thinking in order to get to "know" its essence.
The object is inactive.	The object participates.
Only those characteristics of the living being that can be quantified and measured are studied.	Quantitative and qualitative characteristics are important for understanding a living being.
The living being is studied at a point in time and outside its natural environment.	A living being is studied within its context, natural environment, and evolutionary development.
Living beings are reduced to their individual components.	A living being is seen holistically as being influenced by and in turn influencing its environment. Breaking down the living being into individual components results in a loss of the organism's essence and its integrity.
No special attention is given to the influence of a human centric bias in developing, conducting, and interpreting the research.	The aim is to avoid a human-centric bias and to develop, conduct, and interpret the research from the perspective of the living being.
Difference is the focus.	The focus is on similarities in shapes and processes in nature as evidence of the same forces being in operation.

holistic understanding of the interrelationships that exist in nature and was critical of how they examine only the characteristics that can be quantified and measured at the expense of characteristics that cannot be quantified easily. As a result, Steiner concluded that conventional science cannot provide more than a partial understanding of nature.

Steiner argued that we cannot understand the nature of bees, nor can we assess the effects of beekeeping practices on a bee colony's natural processes, unless we understand a bee colony spiritually. Steiner's spiritual research into the nature of bees led him to conclude that practices such as artificial bee breeding, as well as others that interfere with the natural processes of a bee colony, result in long-term negative effects for their well-being. Steiner stated that the findings presented in his lectures "are not open to doubt, because they are based on true knowledge gained by expending great effort on obtaining these insights" (23).

A Spiritual Understanding
of the Nature of Bees

Life in a beehive is established with extraordinary wisdom behind it. Anyone who has observed life in the beehive will say that. Naturally, you can't say that bees possess a knowledge or science such as human beings do; they simply don't have a brain as we do. They can't take up into their bodies a general understanding of things in this world.
—Rudolf Steiner (1)

Rudolf Steiner started his lecture of February 3, 1923, by answering a question from an expert beekeeper on the difference between bees and wasps. Steiner answered by indicating the importance of understanding the role of cosmic influences in the unique nature of a bee colony:

Cosmic influences have a tremendous effect upon the beehive. You would be able to gain a correct and true understanding of life within the beehive if you were to allow for the fact that everything in the environment that surrounds the Earth in all directions has an extremely strong influence on what goes on in the beehive (1).

Cosmic Forces and the Organization of a Bee Colony

During his bee lectures, Steiner did not explain how cosmic forces affect the Earth, but he did so in his lectures on agriculture, which he presented a year later. In the second of those lectures, he explained how the Earth should be regarded as an organic entity that is affected by the influences of the universe, particularly those of the planets.

The planetary influences do not all work in the same way. Steiner stated that the path of the Sun (as perceived from Earth) divides the planets between inner planets (those closest to the Earth: Mercury, Venus, and the Moon, a planet for this purpose) and outer planets (those farthest from the Earth: Mars Jupiter, and Saturn) (see also Keats, 2009). The inner planets work directly above the Earth, in the Earth's environment. The outer planets work indirectly, and their influences are absorbed by silica in the earth before being radiated back into the cosmos. For the Earth and life on it to function properly, the earthly processes that occur above the ground and the cosmic processes that take place below the ground need to be exchanged continually. This exchange is made possible by clay and calcium. Clay facilitates the upward flow of cosmic influences, while calcium facilitates the downward flow of earthly influences.[1]

Steiner's research found that influences from the various planets explain the differences among insects. Wasps, ants, and bees, for example, have a similar type of social organization, yet, in contrast to wasps and ants, bees are completely under the influence of Venus. As a result, bees have developed what Steiner refers to as "love life" throughout the bee colony. This is a form of unconscious wisdom conveyed in the activities of a bee colony (2).

When observing these activities, one of the first things we notice is the cooperation and success with which bees work together. A bee colony consists of three different groups of bees, often referred to as casts. Each of these different groups has a different role in the colony. The bee colony's fertilized female

---

1. In lecture 6 of his agriculture course, Steiner states there are two kinds of forces that descend from the cosmos: those first absorbed by the Earth before influencing plant life (from Mercury, Venus, and the Moon) and forces that affect plant life above the Earth (from the outer planets).

bee is referred to as the queen bee,[2] and she is responsible for laying eggs.[3] All the other bees in the colony originate from her (or her mother if she has only recently started to lay eggs). The majority of bees in the colony are unfertilized females, and these are referred to as worker bees. They build comb, feed the young bees, clean the hive, and collect and store nectar and pollen. The third group of bees are the male bees or drones; one of their key roles is to fertilize a newly hatched queen bee of their own or another bee colony. Not only do the different members of a bee colony have different roles in the reproduction of the colony, but they also look different.

Steiner attributes the cooperation among these different bees, in particular between the queen and worker bees, to the fact that any desire of individual female bees to procreate or mate is largely suppressed in all but the queen bee. Suppression of the need for sexual reproduction is attributed to the influence of Venus, the planet traditionally associated with love (2). Venus affects certain organs through the bee colony's soul, or astral element. This is one of the four spiritual dimensions Steiner identified as characteristic of living beings.[4]

---

2. Prior to the discovery that the bees in a colony are the offspring of one fertilized female, it was assumed that the role of the larger bee was similar to that of a king in human feudal society. Once it was understood that this bee was female, the name changed from king to queen. However, the term *queen* is not appropriate, as she does not "rule" the bee colony. The relationships among the different types of bees in a colony is much more egalitarian and interconnected and cannot be compared socially with a human monarchy.

3. Worker bees have been seen to lay unfertilized eggs, though they generally do not hatch and are destroyed by the other bees in the colony.

4. The other spiritual dimensions Steiner identified are: mineral (physical dimension); etheric (life force); astral (soul, or consciousness); the "I" (spiritual). He also identified spiritual entities without a physical dimension. Innate objects and materials have only a mineral dimension, whereas plants have mineral and etheric dimensions. The etheric dimension gives plants and other life forms life. Animals have, in addition to mineral and etheric dimensions, an astral dimension, which provides them with a certain level of consciousness. Plants lack an astral dimension but receive astral influences through their associations and interactions with animals, especially insects. In addition to the first

Queen, worker bee, and drone.

In many ways the bees renounce love, and thereby this love develops within the entire beehive. You'll begin to understand the life of bees once you're clear about the fact that the bee lives as if it were in an atmosphere pervaded thoroughly by love.... The bee, with the exception of the queen bee, is a being that would say, if I may put it this way: "As individuals we want to renounce all sexual life, so that we make each one of us into a supporter of the hive's love life." (3)

The love life that exists throughout a bee colony can best be compared with the human experience of "unconditional love."

It is important to realize none of the groups of bees can survive without the others. Nor is reproduction possible unless all groups are present in the colony. The various roles of the female members in a bee colony's reproductive process can be compared to the female reproductive organs in mammals. In a bee colony, the queen bee has powers similar to that of ovaries, namely the production of eggs, while the worker bees perform the role of the womb, protecting and nurturing the eggs as they develop into mature bees.

---

three dimension, human beings have an additional dimension, the "I." Emotions are "located" in the soul aspects of the human "I." Individual animals do not have souls, but share a soul as a group.

The Influences of Venus and Bee Nutrition

The influence of Venus also becomes evident from the different ways in which bees, wasps, and ants obtain nutrition. Bees derive all their nutrition from the nectar and pollen of flowers, which are associated most strongly with Venus and are responsible for plants reproduction, their love life. Wasps and ants do not derive their nutrition from flowers, but from other sources such as the leaves and stem and from animal sources. This interconnection of Venus, the reproduction of plants, and bees means that, by collecting nectar from the plants, bees bring the spiritual forces of the plants' love life into their colony.

The Sun's Influences and the Development of Bees

Steiner's research shows how the reproduction, appearance, and role of the different bees in a colony are the result of variation in the shape of the comb cells and the duration of the bees development from egg to mature bee. A queen bee takes sixteen days to develop from egg to full maturity, in contrast to twenty-one days for a worker bee. A drone, however, takes longer than both and develops in twenty-three to twenty-four days. The following table compares the developmental stages of the queen, workers, and drones:

BEE DEVELOPMENT: DAYS FROM EGG TO MATURITY

	EGG	ROUND LARVAE	STRETCHED LARVAE	PRE-PUPA	PUPA
Queen:	1–3	3–7	7–10	10–11	11–16
Worker:	1–3	3–7	7–10	10–13	13–21
Drone:	1–3	3–10	10–12	12–15	15–24

These differences are no coincidence, but relate to the duration of the Sun's rotation on its axis. Steiner explains that differences in the time it takes for living beings to develop are significant, as this affects the processes that occur during those periods (10).

The twenty-one days it takes a worker bee to develop from an egg to a fully grown bee is the same number of days it takes for the Sun to rotate once on its axis. The result is that the worker bee is exposed to all the influences of the Sun during its development. These influences have been taken up and become part of the worker bee. If the development of a worker bee were to continue beyond this period, they would simply receive more of the same of the Sun's influences (8).

As the Sun comes closer to the end of its cycle, the influences of the Earth on the developing bee become stronger. As worker bees mature at twenty-one days, they are exposed to the Earth's influences as fully developed animals. Thus, worker bees are very much Sun animals that have been exposed to only a little of the Earth's influences just before they hatch (9).

Exposure to the Sun's and Earth's influences is different for drones. The drones have not yet reached maturity at twenty-one days and continue their development beyond the duration of one rotation of the Sun on its axis. As a result, the Earth's forces do influence drones during their development. The extended duration of the drones' development also means that they are further removed from the larval stage than both the worker bees and queen bee (9). Steiner found that the duration of the drones' development from egg to mature bee, together with their increased exposure to the influences of the Earth, gives the drones their male fertility (103).

Queen bees develop in sixteen days; as such, they complete their entire development within the period of one solar rotation. Hence, they are not exposed to the full effects of the Sun's influences, nor are they exposed to the Earth's forces. The shorter duration of the queen's development also means that she remains closer to the larval stage than do the other bees. Steiner found that this enables queen bees to mate and lay fertilized eggs; it gives them their fertility (103). Thus, while all bees receive the influence of the Sun during their development, the queen bees receive no influences from the Earth; the worker bees receive only a little during the final stages of their development; while the drones are affected the most by these influences.

	QUEEN BEE	WORKER BEE	DRONE
Sex:	female	unfertilized female	male
Duration from egg to maturity:	16 days	21 days	23–24 days
Sun's influences:	+	+ — –	–
Earth's influences:	–	– — +	+

The Shape of Comb Cells and the Development of Bees

In addition to differences that are the result of the variation in the Sun's influences during the bees' development, Steiner's research also shows how the variation in the shape of the comb cells contributes to these differences. Steiner points out:

> The most remarkable thing about a bee is not that it produces honey, but that it can produce completely out of itself, out of its own body, these wonderfully constructed honeycomb cells. It has to carry into the beehive whatever it uses as building materials. It works in such a way that it doesn't use this material in the original form, but rather brings it into the beehive already completely transformed. The bee does this all by itself, and what it creates comes directly out of itself. (141)

The different types of bees or castes in a bee colony develop in comb cells of different shapes. Worker bees and drones develop in hexagonal honeycomb cells, with the drone cells slightly larger than those of the worker bees. The cells of queen bees, however, have an elongated spherical shape. The cells in which workers and drones develop are oriented in a horizontal position, while those of the queen bee generally hang down in a more vertical orientation.

	QUEEN BEE	WORKER BEE	DRONE
Sex:	female	unfertilized female	male
Cell shape:	spherical	hexagonal	hexagonal

The reason generally given to explain the shape of honeycomb cells is that six-sided cells offer the most efficient use of space for storing honey and brood. As mentioned earlier, however, Steiner points out that forms and shapes in nature exert certain forces, and that we need to be aware of their effects. With respect to the hexagonal shape of the comb cells, Steiner indicates that this shape is similar to that of quartz, which consists of silica. As mentioned, silica is the mineral substance that captures, transmits, and radiates back the influences of the outer planets from below the earth. The similarity in shape between quartz and the hexagonal cells of the honeycomb indicates that the six-sided cells and quartz are formed by the same spiritual forces. The six-sided honeycomb cells thus also enable the gathering of those influences from the outer planets. Steiner states that this is not a physical force but spiritual, and that it cannot, therefore, be proved using physical methods (183).

The Sun's influences on a bee that develops in a hexagonal cell will be quite different from its influences on a bee that develops in a spherical cell. The shape of the six-sided cells is internalized by the worker and drone larva during their development. By having internalized this energy, the worker bee is in turn able to produce hexagonal cells. Steiner found that the energies embodied in the six-sided cell also enable the bee to perform all their other activities (7).

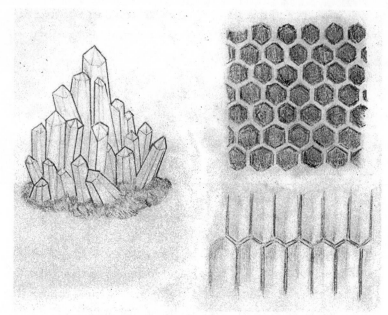

Quartz crystal and honeycomb structure.

The spherical shape of the queen cells prevents queen bees from gathering and internalizing the Earth's influences during their development. Instead, the spherical shape enables the queen bees to remain completely under the influences of the Sun during their development. This is in contrast with workers and drones, which develop in hexagonal cells that are affected by the forces of the Earth (15). To enable queen bee eggs to develop at a faster rate than the other eggs, they are fed differently from workers and drones (103).

When examined from a conventional, materialistic perspective, the queen bees and worker bees have two sets of chromosomes, one from each parent. The drones, on the other hand, develop from unfertilized eggs and have only one set of chromosomes—that of the mother.[5] The genetic makeup of

5. The process whereby drones are produced from unfertilized eggs is called parthenogenesis. Interestingly, the female Varroa destructor mites that are creating havoc to bee colonies throughout the world are also able to produce male offspring without being fertilized.

drones comes entirely from their mother, the queen bee, while for worker bees the queen bee provides only half of their genetic makeup, the other half of which comes from the drones with which the queen has mated.

	QUEEN BEE	WORKER BEE	DRONE
Sex:	female	unfertilized female	male
Sets of chromosomes:	2 (xx)	2 (xx)	1 (x-)

Based on the size of the cell in which the egg will be deposited, a queen bee determines whether she will fertilize an egg. The queen bee measures the cell size by inserting her upper body into it.

It could also be said that all bees are essentially one sex. The queen bees and worker bees are more alike because they have two sets of chromosomes, whereas drones are less so because they only have one set.[6] The degree to which a bee's female characteristics develop is the result of the variation in its exposure to Sun and Earth forces during its development. Queen bees are influenced by Sun forces but not by Earth forces. Worker bees are influenced slightly more by Earth forces and develop in hexagonal cells. Therefore, they do not reproduce sexually. Drones are greatly influenced by Earth forces as a result of the duration and the hexagonal cell shape in which they develop. Hence, they are the least "female."

Steiner makes the interesting observation that a bee colony spends more time and resources on the creation of worker bees than it does the queen bee, and even more on drones (8). However he does not expand on the possible reasons for this.

---

6. Queens and worker bees are referred to as *diploid*, having two sets of (16) chromosomes. Drones are referred to as *haploid*, as they have one set of (16) chromosomes (Laidlaw & Page 1997).

Conventionally, drones are seen as having one purpose only: to fertilize a queen bee. As drones do not collect honey, beekeepers do not regard them as making a valuable contribution to the colony and thus manipulate the colony to minimize the number of drones that develop. However, just because beekeepers do not see drones as contributing to the colony, it should not be assumed that they do not perform important functions from the perspective of a colony. Anthropomorphizing their role potentially ignores any importance that drones may have for the well-being of the colony.

It has been suggested (see Warré 1948, for example) that drones play an important role in keeping the brood warm, thereby enabling more worker bees to leave the hive to forage. This notion is supported by the fact that drones leave the hive only during the warmer parts of the day, when the effort needed to keep the brood warm is at a minimum. It may well be that drones also play a role in the creation of "nest scent" (the importance of which will be discussed later). Drones are not expelled from the hive after the colony's queen bee has mated, but only after the nectar flow stops, when worker bees reduce foraging and stay inside the hive. This, too, lends support to the notion that drones, in addition to fertilizing queen bees, fulfill other roles. It is well-known that drones visit other hives, are allowed entry, and are fed by the worker bees of those colonies. This is generally denied to worker bees from other hives unless they carry nectar or pollen. Furthermore, drones from different hives are known to congregate in certain places to await the arrival of a virgin queen bee.

It is interesting to speculate that drones may well play an important role in gathering information about a colony's external environment and communicate this throughout their own and other colonies. It may well be that drones are expelled during winter in areas where bees hibernate because of the fact that the colony's communication with its external environment slows down during this period. Others, too, have linked the drones' role to communication and to perceiving what is happening in the colony's wider physical and spiritual environment. Hauk (2008, 36) for example sees drones as the colony's "sense organ" (see also Weiler 2006). If we do wish to anthropomorphize bees, perhaps a more appropriate way to view drones would be as the "knowledge workers" of the colony. The various factors Steiner identified as responsible for creating the variation among the different groups of bees in a colony is presented in the following table:

	QUEEN BEE	WORKER BEE	DRONE
Sex:	female	unfertilized female	male
Duration from egg to maturity:	16 days	21 days	23–24 days
Sun's influences:	+	+ — −	−
Earth's influences:	−	− — +	+
Cell shape:	spherical	hexagonal	hexagonal
Sets of chromosomes:	2 (xx)	2 (xx)	1 (x-)

Steiner further highlights the strong effects of the Sun's and the Earth's forces on the development of bees by discussing the manipulation of these environmental factors and how it is thus possible to change the type of bee that will hatch. He stated that, by speeding up the hatching period and manipulating the larvae's nutrition, eggs that were initially intended to become drones will become worker bees, albeit slower ones. Similarly, it is possible to manipulate cell shape, nutrition, and duration of development to change eggs intended to become worker bees into ones that will become queen bees. This is a strategy a bee colony uses when the queen bee dies suddenly.

At the time of Steiner's lectures, manipulation of a bee colony's reproduction by turning worker eggs into queen bee eggs was not thought to have much influence on practical bee-keeping. Today, however, this practice has become widespread, and the majority of queen bees are reared from worker eggs. The artificial breeding of bees will be discussed later in more detail.

The processes that influence the development and reproduc-tion of bees and bee colonies show the intricate nature of the organization and interdependency within a bee colony. It is important to realize that neither individual bees nor the differ-ent groups or castes of bees can survive without the colony, and that the survival of a bee colony is possible only when all groups or castes are present.

Cosmic Influences on Reproduction: Mating and Swarming

Steiner also found that cosmic forces influence the unique ways in which a bee colony reproduces. The reproduction of a bee colony is characterized by two processes. The first is aimed at maintaining the existence of a colony by creating new bees to replace individual bees as they die. The other process is the creation of a new colony through swarming. The first process can be compared to the human body's process of continually replacing individual cells. The second process can be compared to human birth.

Under normal circumstances, the lifespan of the average worker bee varies according to the season, from about six weeks in summer to a few months during the winter in areas where colonies go into hibernation. Under natural conditions, a drone lives approximately forty to fifty days, whereas a queen bee lives for a few years before being superseded by a new queen raised by the colony. As discussed previously, for a bee colony to be able to reproduce itself over time it needs to be able to raise worker bees, drones, and queen bees, and this relies on the availability of fertilized eggs (for the worker bees and queen) and unfertilized eggs for the raising of drones. A queen bee will be unable to lay fertilized eggs unless she mates with at least one drone relatively soon after she has hatched.

Steiner shows how the influence of the Sun is evident in the mating behavior of the queen bee. Queen bees mate with a number of drones high up in the open air, usually on days when the Sun shines. As a result of not having fully completed the Sun cycle during her development, the queen bee feels a connection or attraction to the Sun. During the mating flight, she flies as high as possible toward the Sun, followed by drones from her own and other colonies (11–12). Several drones mate with the queen bee before she returns to the hive and begins to lay eggs. Steiner explains how the drones want to merge their earthly element with that of the Sun during the mating flight. The drones compete for the opportunity to fertilize the queen bee, and only those that are strong enough to fly as high as the queen can mate (104–105). Steiner explained how, when a queen bee is fertilized, the Sun element of the queen bee is touched by the earth element of the drones (104).

Swarming

The process whereby one bee colony "gives birth" to another is called "swarming." Mr. Müller, one of the beekeepers in Steiner's audience, mentioned that the first swarms that occur contain the established queen bee accompanied by worker bees, which leave the hive to establish a new colony elsewhere. Subsequent swarms may contain a recently hatched and mated queen bee (31).

Steiner has described the birth of a new bee colony as quite magical:

> If you now look at a swarm of bees, it is, to be sure, visible, but it really looks like the soul of a human being, a soul that is forced to leave its body. This is truly a grand sight,

to see a horde of bees swarming away. In the same way in which the human soul leaves the body....

There's only a slight difference, in that the human soul has never reached the point of developing its powers sufficiently to create these little creatures. Within us there is continually the tendency to want to do this; we want to become many tiny creatures. We actually have within ourselves a tendency to continually want to change the form of things inside our bodies so that they become like crawling bacilli or bacteria, or like little bees, but we suppress this process before it happens. In so doing we become complete human beings and not a collection of tiny entities. But the beehive is not a complete human being. Bees can't find their way into the world of spirit. We have to take them to an empty beehive to bring about a rebirth, or reincarnation, of these bees. This is very directly a picture of reincarnating human beings. That person who can observe such a phenomenon develops a tremendous respect for these old swarming bees with their queen, a swarm that acts the way it does because it is trying to find its way into the spiritual world. But this swarm has become so caught up in the material world, so physical, that it can no longer fulfill this desire. So the bees cuddle up together and become a single body. They want very much to be together. They want to leave this world. And you also know that while they are flying about, they might attach themselves to a tree trunk or something similar, and then they press themselves together in order to disappear, because they want to gain entrance to the world of spirit. And then they once again become a proper colony when we help them by bringing them back to their new beehive.

So, from this description you could probably say that insects teach us nothing short of what might be considered the highest understanding we can derive from nature (156–157).

Steiner explains the process that leads to swarming as follows: Bees have five eyes—two on the side and three that are very small on top of their heads. Bees see very little with their eyes and live in a kind of twilight. Thus they navigate by a sense somewhat between smell and taste. This twilight forms a barrier that closes a bee off from its external environment. A bee's poor vision is a result of the absorption of tiny amounts of bee venom. The bee venom enables the bee to carry this barrier with it when it leaves the colony.

Once a new queen bee has hatched, the Sun's influence enters the hive through a new queen. As the new queen bee has developed completely under the influence of the Sun, she brings a new source of "sunlight" into the beehive. The result is that the tiny eyes on the worker bees suddenly become active; they become "clairvoyant," as Steiner calls it, and have an aversion to the light coming from the new queen bee (13–14).

The effect of the new queen is that the bee venom of the old queen is disturbed and can no longer contain the colony within itself. This effect of bee venom, Steiner found, occurs every time a bee colony is disturbed by an external influence, a bee colony wants to remain undisturbed and completely self-contained (15). The disruption of the bee venom raises the fear that the bees will no longer possess the bee venom needed to defend the colony, and in response they move away from this source and swarm, leaving the hive (156). The bees that remain behind are born under other conditions and do not experience this tension or fear in the same manner (14).

During the lecture, Müller commented that, in addition to a swarm occurring after a new queen bee has hatched as described by Steiner, swarming can also occur nine days before a new queen bee has hatched. Steiner clarified his earlier comments and stated that the egg of a future queen bee gives off the most light after seven days, or nine days before the new queen bee hatches, and, in fact, the strongest radiation appears when the queen bee is still in the larval stage (33).

Steiner explained that it is not the light as such that affects the bee venom, but the chemical reaction resulting from the light. For bees, the appearance of a new queen may be compared to our view of the phosphorescent-like light of glow worms or fireflies on a summer night (30). The effect of a new queen on the bees' eyes is not the same as our sense of seeing, as bee eyes do not have the light sensibility of people's eyes (30). Steiner stated that it is inappropriate to assume that bees experience the world as humans do simply because they have eyes. Again, we have to be very careful not to anthropomorphize nature:

> You would never reach the point of being able to imagine the whole matter in any way other than by saying that, because the human being sees things this way, the animal, too, must see them the same way. (35)

Steiner stated that bees are supersensitive to chemicals, and that, for bees and other animals, smell and taste are more important than sight and hearing. Hence, it is important to explore these chemical senses when studying nature (37).

Similarities between a Bee Colony and the Human Body

Steiner also showed that, in addition to the creation of inter-connectedness of the bees in a colony through the influences of the Sun and Venus, further evidence of this level of interconnectedness is found by researching the similarities between the bee colony and the human being. As discussed, the similarities found in nature are evidence of the underlying spiritual forces that created them, since the spiritual dimension has primacy over the material dimension. Steiner saw the human being as a microcosm, and, as such, the relationships among the different organs of a human body are replicated throughout nature. By looking at similarities from this perspective, all the processes of nature are symbols and representations of what occurs in human life (159).

With respect to a bee colony, Steiner pointed out that the same processes happens in a bee colony and in the human head and, as a result, have similar capacities (16). The relationship between the queen, workers, and drones within a bee colony is similar to the relationship that exists among the protein, blood, and nerves cells.

Steiner found that the development and roles of the human nerve cells, blood cells, and protein cells are similar to the development and roles of the queen bee, workers, and drones in a colony. The individual nerve cells in the human head take the longest time to develop, and if they could develop in all directions as in the bee colony, "these nerve cells would become drones" (16). Like drones in a bee colony, human nerves are constantly worn out by the human body. However, unlike drones, the human body does not create and destroy its nerves on an annual basis, as this would lead to death for the body. However, Steiner did say that the weakening of our nerves means that we are decreasingly able to sense our bodies, and that once we have frayed our nerves and worn them out, we die (17–18).

Although Steiner did not make the connection, the similarities in function between our nerve cells and drones also supports the notion that the role of drones is to transfer information both within the colony and from the external environment.

Steiner showed that, if human blood cells could become little animals, they would be like the worker bees, which collect nectar and pollen from plants and use it to produce wax for the honeycomb cells. Similarly, the blood cells move from the head throughout the body, transporting nutrients. The bones built up by the blood cells also have a shape similar to the six-sided honeycomb cells The other cells of the body, muscles for example, also show characteristics similar to the shape of honeycomb cells; however, in these other cells of the body, this shape is less solid.

Steiner concluded that human blood cells and the worker bees contain the same forces (17). The similarities between worker bees and blood cells can be further expanded, since blood cells, too, defend the body against interference from foreign organisms such as bacteria and viruses.

Baker (1948) applies Steiner's spiritual approach to understanding the bee colony by expanding the similarities Steiner identified between plants and insects, and particularly between the *Hymenoptera* (sawflies, wasps, bees, and ants) and trees. Baker comments that awareness of the similarities between trees and bees enables a better understanding of the interconnectedness within a bee colony and that such understanding leads to greater success in the practice of beekeeping.

Steiner observed that both plants and bees undergo a series of metamorphoses, from seed to plant to flower and from egg to larva to pupa to bee. Trees are bound to the earth, while bees are free.[7] He pointed out that both trees and bee colonies construct cellular bodies, which they need to function. The cambium of a tree forms its maternal element that ensures it will develop and grow. This is similar to the queen bee, which ensures the growth and development of the colony. Tree leaves transform and carry nutrients into the trunk to enable it to grow. In the same way, the worker bees gather pollen and nectar to feed the bee colony and enable it to grow. Tree blossoms produce pollen, which is carried into the air to fertilize other trees, similar to the way in which drones fly into the air to fertilize the queen bee.

There are also similarities in the seasonal rhythms of deciduous trees and bee colonies. Both expand and grow in summer and contract and rest during winter. Like trees, a bee colony sheds its flowers (drones) and most of its leaves (workers) in winter, so that only its trunk (the queen bee and a small number of worker bees) remains in winter.

---

7. Steiner found that, during the earlier stages of their evolution, plants had a cloud-like consistency and were not as solid and bound to the earth as they are today.

Steiner also showed the effects of the interconnectedness within a bee colony in his explanation of the relationship between a bee colony and the beekeeper. This came in response to a comment from his audience about a belief among beekeepers in traditional farming communities that there is a kind of soul connection, or spiritual bond, between the beekeeper and the bees. The belief gave rise to the practice that, when the beekeeper dies, this must be communicated immediately to the bees, and that failure to do so results in the death of the bee colonies during the following year. Mr. Müller confirmed that there is some evidence to support this phenomenon. Steiner stated that the way bees respond to the mood of a beekeeper is evidence of a spiritual connection between the beekeeper and the bee colony. If a beekeeper attends to the bees while in an angry mood, for example, the bees are more likely to sting than when the beekeeper is calm and works in harmony with the environment (59).

Steiner further explained this interconnectedness by explaining that the bee colony as a whole recognizes the beekeeper, not individual bees. He compares this process with the human ability to recognize someone who has been away for an extended period of time—long enough for all the cells in the human body and that of the former acquaintance to be renewed (67). In other words, there is no individual cell in either body that has had a direct experience of knowing the person before, yet one's ability to recognize the other has been stored by the human body as a whole, not by the individual cells. The hive's ability to recognize the beekeeper indicates a similar process. Steiner states:

> The beehive isn't simply what you would call a collection of an undetermined number of bees, but rather a complete whole, a complete being. That which you can recognize again, or not recognize again, is the entire beehive....
>
> With bees, this is the very thing you must pay attention to: that the object of our concern has nothing to do with

the individual bees but rather with that which absolutely belongs together to make up the whole. (68–69)

The fact that a bee colony forms an individual organism and not a collective of individual bees means that the whole colony has an astral dimension, a consciousness similar to individual organisms. Steiner concludes, "You need to study the life of bees from the standpoint of the soul (3).

Steiner's research shows that it is important to see a bee colony as a whole and that this is why we can learn so much from a bee colony; it completely contradicts the conventional understanding we have formed. He states:

> You will get on well with bees only if you go beyond the normal, basic understanding of things and actually begin to follow matters with an inner eye. The picture of things you get in this way is indeed wonderful. Using this type of insight, you'll have to say that a beehive is a total entity. You must try to understand it in its totality. (73–74)

Discussion: Understanding a Bee Colony

Steiner's starts holistically in his spiritual research on the nature of bees, examining the impact of the influences of Venus on the way intricate relationships have evolved among the different types of bees within the colony and by focusing on the similarities between a bee colony and the human body. In this, he differs fundamentally from conventional research on the nature of bees.

Steiner explained how planetary influences are reflected in the bees' social organization and reproduction. The influence of Venus on the development of bees is also seen in the relationship between bees and plants. Bees derive all their nutrition from flowers, the plant's organs for sexual reproduction, which are strongly associated with Venus. Through bees, the influences of Venus find their way from plants into honey and into human beings who consume the honey.

Steiner showed, too, how the variations among the different bees in a colony arise from the shapes of the comb cells and how time mediates both the spiritual and physical influences as bees develop from egg to maturity. His research shows how, by manipulating these factors, it is possible to alter an egg intended to develop into a worker bee to become instead a queen bee. This has become a standard beekeeping practice.

Steiner further elaborated on the strength of the interrelationships among the different bees in a colony by identifying the similarities between a bee colony and the human body. He discussed the roles of the queen bee, worker bees, and drones in a colony and how those functions are similar to those of the protein, blood, and nerve cells in the human head. He showed how the bee colony's construction of honeycomb is similar to the way the human body is built. Steiner continues by pointing out the similarities among the different groups or castes of bees in a colony and different cell types in the human body by examining the time it takes for each to develop. The protein cells develop fastest in a human embryo, followed by the blood cells, while the nerve cells develop last. The same occurs with the development of the queen bee, worker bees, and drones in a colony. These similarities show that the forces that shape the intricate relationships between the different cells in the human body are also evident in the bee colony.

This interconnectedness of a bee colony is seen not only in the relationships within a colony, but also in the colony's interaction with its external environment. Similar to blood cells, which defend the human body against foreign bodies that have entered it, the worker bees of the colony, too, respond to interference with defensive behavior involving bee venom. The difference is that, unlike the human body, a bee colony does not construct its own skin to help protect its internal organs from environmental influences, but depends on its environment for this.

Throughout Western history, the relationships among the individual bees in a bee colony have been seen as analogous to human society. Steiner shows, however, that it is more appropriate to compare individual bees with the cells of a human body and the different types of bees with human organs. His research led him to conclude that, to understand the unique nature of a bee colony, it needs to be seen as an organism rather than as a collection of individual bees.

4

SPIRITUAL RELATIONSHIPS:
BEES, PLANTS, HONEY, AND PEOPLE

"You see, these bees, wasps, and ants are not simply rob-
bers; at the same time, they bring to the flowers the possibility of remaining alive." (134–135)

"Bees wouldn't have been able to sit on flowers for the past
few thousand years if they hadn't cultivated them by nib-
bling at them to keep them healthy." (136)

"Honey is the form of nutrition that must have the most
salutary and beneficial effects on a human being." (53)

During Steiner's lecture of December 12, 1923, Mr. Dollinger asked him about the relationship between bees and
flowers. "He wants to know what kind of relationship it is that
ties them together and what importance honey has for human
beings" (103). Steiner pointed out that, to understand the relationship between bees and flowers, we need to acquire a holistic
perspective with a focus on the interrelationships between bees
and plants on the spiritual level:

Around us are not only elements such as oxygen and
nitrogen, but there is present throughout nature also
intelligence and understanding. No one is astonished if
you say, "We inhale the air." ...[but] consider it a bit of
fantasy if you say that, in the same way we inhale air

with the help of our lungs, we also "inhale" this power
of mind, for example, with our noses or ears. (140)

As mentioned, Steiner rejects one-sided perspectives and linear approaches to studying nature, as these break down complex
relationships and examine individual components apart from
their context. He illustrates the limitations of such a perspective
by pointing out that bees collect nectar and pollen and transform
them into honey and wax, which have value for human beings.
From a human perspective, bees are therefore considered useful.
However, from the plants' perspective, bees could be seen as rob-
bing them of the nectar and pollen they need for their reproduction. From the perspective of the flowers, they would be better off
without bees. Neither of these approaches however provides a true
picture of the relationship between bees and flowers (126–127).

Steiner rejects the notion that the driving force behind nature
is competition and survival of the fittest, ideas that largely domi-
nate the materialistic understanding of biological evolution.
Instead, Steiner sees the relationship between bees and flowers
as one of cooperation and mutual benefit.

Much like the relationships in a bee colony, the relationship between bees and plants is strongly shaped by cosmic
influences. The bee lectures do not go into much detail about
how these forces influence plant growth, but Steiner discussed
this in 1923 in lectures on elemental beings[1] and in his lectures on agriculture in 1924. The information presented here
on plant growth and on the role of bees in pollination is taken
from the material Steiner presented in those lectures.

Understanding Plant Growth from the
Perspective of Spiritual Science

As discussed earlier, the Earth and its immediate atmosphere can be regarded as an organic entity, an organism that

_____

1. See Rudolf Steiner, *Nature Spirits, Selected Lectures* (Rudolf Steiner
Press, 1995).

is influenced by forces coming from the universe. By examining the Earth in the context of its evolutionary development, Steiner found that, in the current phase of its evolution, the Earth is very slowly dying, and life on it is increasingly losing vitality; it could be said to be getting sicker and sicker. During Steiner's time, this loss of vitality was seen as being hastened by the use of chemical fertilizers. His lectures on agriculture seek to restore the earth's loss of vitality through biodynamic agricultural practices.

It was noted earlier that the Earth and every physical entity on it are expressions of one or more of four spiritual dimensions. Inanimate objects and materials have only a mineral dimension, whereas plants have a mineral and etheric dimension. The etheric aspect gives plants and other living forms the animation of "life." Plants receive astral influences through their associations with animals, especially insects. Steiner found that, because plants are comprised only of the mineral and etheric dimensions, they are affected strongly by influences emanating from the planets and the minerals substances silica and calcium. Silica stimulates plants to absorb the influences from the outer planets and capture these in the soil. Calcium awakens plants to the influences that come from the inner planets. Silica gives plants shape and form and enables them to counteract the forces of gravity. In addition, silica is responsible for the nutritional qualities of plants, whereas calcium supports a plant in its ability to reproduce.

Etheric Formative Forces

With respect to plant growth, the etheric dimension, which gives plants life, is expressed through four etheric formative forces: heat, light, sound (or chemical), and life. These etheric formative forces have a number of unique characteristics. Each corresponds to conditions referred to as the four traditional elements: fire, air, water, and earth. Conventional science uses the terms *heat, gas,*

ETHERIC LIFE FORCE	PLANT	ELEMENT	ELEMENTAL BEINGS AND MYTHOLOGICAL DEPICTIONS	FUNCTION FOR PLANT	ANIMAL SPECIES	EMOTION OF THE ELEMENT
life	roots	earth (solid) / water	earth elemental beings (gnomes)	Feeds the roots and thrusts the plant out of the earth.	frogs and toads	dislike
sound / chemical	leaves / stem	water (fluid) / air	water elemental beings (undines)	Transforms the cosmic information brought up through the roots; carries the action of the chemical ether into the leaves.	fish	sensitivity
light	leaves / stem / flowers	air (gas) / fire	air elemental beings (sylphs)	Shapes the plant out of the chemical ether.	birds	love
heat	fruit / seed	fire (heat)	fire elemental beings (salamanders, or fire spirits)	Carries the action of heat into the seed through the transfer of pollen.	insects (butterflies and bees)	sacrifice

Compiled from Rudolf Steiner, Nature Spirits: Selected Lectures *and* Agriculture: Spiritual Foundations for the Renewal of Agriculture (*also published as* Agriculture Course).

liquid, and *solid.* Steiner found that these etheric formative forces are associated with "elemental beings." He also noted that such beings are depicted in northern and central European mythology and that they embody particular feeling states (see the table). Elemental beings are also associated with certain animal species from which plants derive astral influences. The qualities and characteristics of the etheric formative forces and elemental beings are summarized in the table.

With respect to plant growth, the life etheric formative life force works through earth elementals, which are responsible for pushing information and energy up from the earth through plant roots, thus pushing the plant up and out of the soil. At the material level, they enable plant roots to take nutrients from the soil and into the plant. earth elementals are associated with toads and frogs and characterized by a certain degree of antipathy or dislike. In their tasks for plant growth, they join forces with the water elementals.

Water elemental beings are associated with the chemical etheric formative forces. They are responsible for transforming the information and energy that flows up from the roots with the downward flow of terrestrial information and energy from the leaves and stem, as this is created by the air elemental beings. This is expressed at the material level in the chemical process of photosynthesis. The water elementals have sensitivity toward fish.

The air elemental beings are associated with the etheric formative forces and convey light to plants. They create and form the spiritual archetypical shape of the plant according to the information and energy flowing up from the water elementals. This plant information trickles down into the soil when a plant withers at the end of the summer. The archetype is perceived in the soil by the life etheric formative forces. Moreover, the air elementals are receptive to movements and vibrations in the air produced by birds and insects. A feeling of deep sympathy, or love, exists between air elementals and birds.

The fire elemental beings, which are linked to the fire etheric formative forces, transmit warmth to the pollen and the seed carpel during pollination. Warmth from the fire elementals enables seeds to develop and ripen. Once a seed has ripened and is planted in the earth, the warmth of the seed combines with the archetypal plant information contained in the life etheric formative forces in the soil, which give form to a new plant. The fire elementals play an important role in plant reproduction and survival. The fire elementals have a close affinity to the insect world, in particular to bees and butterflies. They greatly enjoy following these insects as they fly from flower to flower.

Steiner found that each insect has a specific aura, and that the presence of the aura is especially strong in bees, intensifying as they suck nectar and carry pollen from one flower to another.[2] The elemental beings associated with the air and fire etheric formative forces in particular are strongly attracted to this aura, and the fire elementals feel so closely related to the bee that they want to unite with the insect. The aura arises when bees pollinate flowers and the fire elemental sacrifices itself as it transfers the warmth from the fire etheric formative force onto the seed carpel. As a result of this sacrifice, the plant is able to form and ripen seed and complete its growth cycle (173–174).

At the material level, when bees and the fire elementals pollinate plants, miniscule amounts of bee venom are transferred from the bee and absorbed by the plant. Bee venom has great curative powers, as do insect venoms in general, and ant venom (formic acid) in particular. The therapeutic powers of these venoms rests in their ability to revitalize plants. Without bees visiting flowers to take nectar and pollen, bee venom would not be transferred to the flowers. The result would be that, over time, plants would lose their vitality and die out (134). The revitalization of plant life that results from this relationship between bees and flowers also occurs by wasps and ants at other part of the plant. When wasps and ants feed on the plant juices in the leaves and stem, the juices are exchanged with wasp and ant venom, which revitalizes the plant and averts the dying process

2. In a 1908 lecture, Steiner linked the etheric aura that arises when a bee is taking nectar from a flower and, more particularly, with a swarm of bees that has landed in a tree and takes off again with the presence of sylphs and lemurs (undines) (Steiner 1998, 175). In a 1923 lecture, Steiner attributes the aura surrounding a bee to the presence of fire spirits (see Steiner 1995, 129).

(135). This is why Steiner found there is such a deep relationship between insects and plants, bees and flowers (137). It could be said that bee venom has a similar role in revitalizing plant life as the preparations used in biodynamic agriculture have for the soil.

Steiner shows that the mutual benefits of the relationship between bees and plants is also evident from the fact that fruit trees and similar plants thrive better in areas where bees are kept than in areas without bees. This is true despite the fact that bees take pollen that the plants need to reproduce (20). At first glance, it appears as though bees remove the nectar and pollen that plants need to form fruit and procreate, but on closer examination, we see that bees enable plants to be revitalized and continue living (139). Studying the relationship between bees and flowers from a spiritual-scientific perspective, we recognize the existence of mutually beneficial relationships both for bees and for the continuation of plant life on Earth (154). In addition to the benefits that human beings derive from healthy plant life and the pollination of food crops by bees and other insects, bees also produce an important food, honey.

Honey and People

Rudolf Steiner responded to the second part of Dollinger's question on the benefits of honey by discussing the inadequacy of conventional nutritional research that claims the benefits of foods can be discovered by breaking these down into their chemical components and testing for nutrients. Steiner argued that, to understand the effects of food on the human body, it is necessary to focus on food holistically and, in particular, that the connections less obvious to a conventional understanding of nature are really the most important.

In relation to honey, Steiner pointed out that nectar and pollen are substances prepared by nature to be extraordinarily refined and, as a result, they are rare and difficult to obtain. If not for bees, people would consume very little honey, as it is distributed in very small quantities among the plants (5–6, 18). Steiner showed that, to understand the true benefits of honey for human beings, we need to realize that bees obtain all their nutrition from the flower nectar and pollen, the part of the plant most strongly associated with Venus. This means that the influences of Venus find their way from the plant's nectar into the bees' honey and into human beings through the consumption of honey.

In addition to the influence of Venus, the spiritual qualities of honey are also created by the activities of elemental beings. As we've seen, flowering is directed at the spiritual level by the elemental beings associated with the etheric formative forces of air and fire. As bees pollinate flowers, an aura is created as the fire elementals sacrifice themselves. This is experienced by bees as a kind of love. Steiner states that this feeling of love is also absorbed into the nectar the bees carry back to their hives and into the honey (3).

According to Steiner, honey is beneficial to us because the astral forces of honey regulate the body's organs. The nectar that bees collect is a plant product and thus has no astral dimension; but when nectar is transformed into honey, it is enriched with astral forces by the bees. Through the astral element in honey, people learn to work on their internal organs and to restore the effort needed by the soul to become sound. The astral force in honey establishes the proper interconnections between a person's elemental forces. Steiner stated that there is nothing better for people than to add a little honey to their food (3–4).

Beneficial effects of honey also result from the hexagonal cells in which it is stored. Steiner explained that bees collect pollen and nectar by locating flowers through a kind of scent-taste rather than through sight. Worker bees can "taste" the nectar and pollen as they fly toward the flowers (12–13).[3] Once

---

3. Zoology professor Karl von Frisch (1954) showed that bees communicate the location of pollen and nectar sources to the colony through the "waggle

they reach a flower, they secure themselves on a flower by means of their claws, and then draw up the nectar and collect pollen. Pollen are deposited and carried on the "brushes" on the bee's hind legs. The flower nectar is drawn up through the bee's proboscis and deposited mostly in a honey sack within the bee's body. After the bee has gathered enough nectar, it returns to the hive and regurgitates it to feed other worker bees, which then deposit it as honey in the hexagonal comb cells (12).

As discussed earlier, the forces that create the shape of silica and quartz that enables these to capture and transmit influences from the outer planets into the Earth's surface also create the shape of the comb cells. Hence these cells, too, are able to capture and transmit the influences from the outer planets. As honey is stored in six-sided cells and is produced by the bees that developed in those cells, it, too, has absorbed these influences. As a result, the honey is strongly enriched by both etheric and astral influences. By eating honey, we take in these forces (177).

Steiner explained how the consumption of honey affects the human body by referring to the similarities between it and the bee colony. The force responsible for the shape of quartz and silica, substances that provide shape and form to plants, also shapes the hexagonal cells. When bees collect nectar, they take the forces from the flower into their bodies, which enables them to build the six-sided wax comb cells (183). Steiner stated that the construction of cells from wax is similar to the way human bodies are formed, and that the human body consists of and needs six-sided spaces. It is not easy to show that the human body is made by blood cells from a kind of wax, but Steiner found that this is nonetheless a fact (17). In the human body, this substance exists in a flowing, liquid form. Similar to what happens below and above the earth, there is a continuous flow of quartz or silicic

dance." See Karl von Frisch, *The Dance Language and Orientation of Bees* (Cambridge, MA: Harvard University Press, 1993).

acid flowing from the human head into the rest of the body. In the human body, however, the quartz does not crystallize (49).

Thus, if we eat honey, it provides us the same forces that enable bees to construct honeycombs, (177) the strengthening force responsible for the creation of shapes and forms. Honey contains the power to maintain the form and solidity of the human body (19). Eating honey can make up for the loss of this force by producing solidness when people become too weak to develop it themselves (18–19).

Steiner found that this extraordinarily beneficial effect of honey becomes particularly important when people age, as honey assists in maintaining the form of our bodies. He cautions against eating too much honey, however, which will result in too much form, making the body brittle and prone to diseases. A healthy person will have a natural sense of how much to consume (18–19). Steiner asked,

> What is it really that brings about the effect that honey has on human beings? It is the hexagonal formative force that works on the human being through the honey. This power is inside the bee. You can see its effects by observing the beeswax cells. It is through this power that honey can be of such benefit to humans. (54)

Cosmic Influences on the Availability of Nectar

During his lecture of December 1, 1923, this question arose:

> There is an old farmer's saying that states that if it rains on the third of May, the nectar gets washed out of the blossom on flowers and trees so that honey production is reduced that year. My observations during the past four years seem to indicate there may be some truth to this adage. Is something like this really possible? (61)

Steiner explained that it is not so much the exact date that matters, but the season. What is important is the specific area of the sky from which the Sun's forces are coming, not whether the Sun shines or not. He points out that from every corner of the sky the Sun affects the Earth in different ways. It is not the Sun itself that has these effects, but the constellation of the zodiac behind the Sun, which assimilates the influences of the zodiac and passes them on in its rays.

Steiner explained that, when the Sun is in the Aries region of the sky, it can bring its full force to bear upon flowers, which develop the substances found in nectar. However, when it rains during this period, the earth forces are greater and minimize the Sun's effects. As a result, the flowers are unable to develop while under these influences, but will develop during a later Sun period, or flowering will be interrupted altogether and the bees will not find any nectar (62). As the Sun moves to the next region of the zodiac (Taurus), sunlight comes from a different area of the sky, which hardens and dries plants rather than stimulate the formation of the flowers, and this makes it more difficult for them to produce nectar.

It is likely that the influences of the zodiac—which affect the amount of nectar in plants and, consequently, the amount of honey produced—also affect the characteristics of the nectar and of the honey. Steiner did not elaborate on how these influences affect honey, nor has the matter received much attention since. This contrasts with the abundant research in response to Steiner's lectures on agriculture and the effects of cosmic influences on plant vitality.

In a subsequent lecture, Steiner was asked,

> With regard to the influence that the signs of the zodiac can have on the production of honey, in some older, more traditional farming communities, farmers still pay close attention to this—for instance, when they sow seeds, they do so when the moon is in Gemini. Are they following the characteristics of the zodiacal signs superficially, or is there something else that might be meaningful? (81).

Steiner stated that, because bees, specifically the queen, are "Sun" animals, the movement of the Sun through the zodiac influences them greatly, and that changes in the Sun's influence and effects are passed on to the bee. He explained that, because bees depend on plants, there is a connection between sowing seeds under a certain sign of the zodiac and the substances bees then find in the plants. However, Steiner also did not elaborate on that relationship. Little attention has been given thus far to how the different times of sowing and planting nectar plants and others considered beneficial to bees affect the nectar collected and the honey produced or the health of the bees.

Steiner pointed out that there is a lot of truth in farmers' traditional knowledge as reflected in their sayings, but that the true meaning of these sayings has unfortunately been lost to be replaced by superstition (63). He suggested that such traditional farming practices should be properly examined scientifically to provide a solid basis for such sayings (82).

Bee Venom

In response to a question from Mr. Müller—how a bee sting can pose such a great danger to certain human beings—Steiner discussed the effects of bee venom on people (67). He explained that bee venom affects our "I." As discussed earlier, human beings are comprised of four spiritual dimensions: physical, etheric, astral, and "I"-being, Steiner said that, while this spiritual dimension (the "I") is not generally found in the mineral world, plants, and individual animals, in a mysterious way the "I" is present in bee venom (106). He explained that, when we are stung by a bee, the venom goes into the body. If we are healthy, the venom will strengthen the "I" by energizing the blood and increasing its movement. The result is irritation and possibly inflammation. However, it will not have a negative effect on the heart of a healthy person. If, however, the heart is diseased, bee venom can strengthen the "I" to a dangerous level. The venom can push the blood too hard against a diseased heart valve and lead to unconsciousness and even death (107).

As Steiner explained, despite the fact that the venoms of bees, wasps, and ants (formic acid) cause inflammation and similar effects, they can also be used therapeutically (137). Venoms generally gather spiritual elements and can thus be used as remedies (134).

Discussion

Steiner indicated the importance of paying attention to the creative processes that shape plant life and to noticing the kinds of relationships in nature that form the basis of natural processes (183). He explained that, given the relationship between bees and flowers and the importance of honey for human beings, we need to examine this holistically by focusing on cooperation and mutual benefit rather than on competition and survival of the fittest.

Thus, Steiner showed the importance and benefits of bees for humanity. Bees aid in the continuation of plant life, both spiritually and physically. Spiritually, bees enable the etheric formative forces to bring warmth to seeds, allowing them to ripen and plants to reproduce. Physically, bees enable plants to remain vital by depositing bee venom in plants during pollination and through their role in the formation of seeds.

By producing honey, bees also benefit people. Honey enables us to derive benefits from the cosmic forces captured by plants in the nectar, the astral forces imparted by bees when nectar is transformed into honey, and the etheric and astral forces from the quartz-shaped comb cells when the honey is stored and ripened.

By studying the similarities between a bee colony and the human body, Steiner also provided other reasons for the importance of honey for humans. He explained that the forces that enable bees to make hexagonal wax honeycomb cells can be used by us to construct our own form (18–19). Such forces are transmitted through silica in nectar and honey from the plant to the bee colony and into the human body. Honey thus strengthens us, both by working on the astral body and by providing form to the physical body.

Despite identifying the influences of cosmic forces on the availability of honey and plant life, Steiner did not elaborate on how these influences affect the quality of honey. Similarly, little attention has been given to the effect of cosmic influences when nectar plants are sown in relation to what the honeybees produce.

The many benefits bees provide led Steiner to conclude that beekeeping can be a great aid to human culture (4).

5

HIVES AND HIVE MANAGEMENT

Rudolf Steiner's research shows that the interactions between bees and plants are part of an integrated system based on cooperation and mutual benefit. Plants provide bees with nutrition, and bees revitalize plant growth at the spiritual and physical levels. We benefit from this relationship by consuming honey and the fruits that result from pollination aided by bees. Human beings clearly benefit by interacting with bees, but the benefits for bees are less obvious.

In his comments on the discussion between Erbsmehl and Müller, Steiner explained that many modern beekeeping practices were developed to meet particular economic objectives. In the twentieth century, they were intended to maximize output and economic returns over, say, the well-being of bee colonies or the quality of honey. Steiner cautioned beekeepers to be careful about introducing beekeeping practices without understanding how they affect the bees' natural processes. This chapter discusses beekeeping practices and their degrees of interference with the bee colony's natural processes as identified by Steiner and others who sought to develop beekeeping practices that work in harmony with the bee colony.

Beehives and Hive Management

A key finding of Steiner's research is that a bee colony should be understood as an organism with internal processes that have been established through a great deal of wisdom. As with any organism, a bee colony dislikes interference with its internal processes, as this disrupts the bee colony's proper functioning. The defensive behavior with which a colony meets any intruder is evidence of this. From this observation arises the key beekeeping principle for working in harmony with the bee colony's natural processes: the need to minimize all types of interferences.

A cursory overview of beekeeping history suggests that, even before the introduction of modern beekeeping practices, the focus of the relationship between bees and people was aimed at realizing human objectives without much concern for the bees. Initially, people hunted for honey by locating wild bee nests in trees or in rock cavities. The little evidence available suggests that the process by which honey was extracted took place with scant regard to the survival of the colony (see Crane 1983 and 1999 for an overview of the history of beekeeping).

With time, hunting for honey gave way to the more permanent practice of tree beekeeping in forested areas. The wild bee colonies became more or less managed and owned by individuals or groups. Tree beekeeping developed to quite an advanced stage, whereby cavities were created in existing trees in which bees could settle. It is not a big step to go from tree beekeeping to creating log hives by taking the part of the trunk in which a colony had settled to a different location. A swarm that had not settled in a tree cavity was captured in a woven cane or wicker basket called a *skep*.

In areas with few trees, such as the arid parts of the Mediterranean, bees made their nests in rock cavities. In these areas people developed hives by imitating the nests built by bees, keeping the colonies in cylindrical hives made from clay or reed covered with mud. Thus two types of hives arose: upright hives made of wood or woven from plant material and horizontal, cylindrical hives made from clay or plant material covered with mud. Over time, both upright and horizontal hive designs developed in a wide range of styles, sizes, and of different materials, depending on their availability.

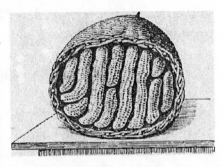

Bottom view of a wicker skep showing the pattern of natural comb building (from Abbé Warré, Beekeeping for All).

With both hive types, honeybee colonies were pretty much left to their own devices throughout the year. The main activity of the beekeeper was taking the honey and wax. The great advantage of the cylindrical hives was the ability to extract the honey without destroying a colony, as the cylinders often had a moveable cover at the back and the entrance for the bees to the front. The extraction of honey from the upright hives was more difficult. The most popular and effective method for taking honey from such hives was to smoke out the hive and subsequently destroy their nest, or to asphyxiate the bees with sulphur. Either strategy virtually guarantees the colony's destruction. The only practice that provided a bee colony with some chance of survival was beating the hive with sticks to force the colony into refuge elsewhere. Depending on whether the queen was able to escape with the colony, it could potentially be re-hived.

In the mid-nineteenth century, a number of changes in beekeeping practices occurred. The most important of these were the discovery of bee space, the introductions of wax foundations, the queen excluder board, and the smoker. In 1851, an American beekeeper, Rev. L. L. Langstroth (1810–1895), figured out how to avoid cutting the honeycombs attached to the walls of the beehives. He realized that bees are fairly consistent in the amount of space they leave between their combs. This came to be called "bee space" and is between 1/4 inches and 3/8 inches (5 mm and 9 mm) (see Bailey 1982; de Bruyn 1997).

The discovery of bee space led to the development of hives that incorporate moveable frames for bees to build their comb. (Hives with moveable frames will be referred to as "framed hives" throughout this book.) Many different designs have been developed on the principle of moveable frames. There is some variation from country to country, but the majority of modern beehives around the world follow variations of the Langstroth, National, or British Standard designs (Bailey 1982; De Bruyn 1997). In general, framed hives consist of one or more oblong boxes with rows of stacked moveable frames that house the brood chamber and stores of honey and pollen. The boxes are placed on a floor board placed either directly on the ground or on a stand. A cover is placed over the frames in the top box, which in turn is covered with a roof. The hives are opened from the top for inspection.

Their size varies depending on the particular design. For example, the internal dimensions of an eight-framed brood box of a Langstroth hive are 18 1/4 inches (456 millimeters) by 12 1/8 inches (303 millimeters), with a depth of 9 1/2 inches (237.5 millimeters) and a volume of approximately 1 1/6 cubic feet (33 liters). As the colony grows in size, additional boxes with empty frames (called "supers") are placed on top of the brood chamber. The supers are generally shallower than the brood chamber and thus also have a smaller volume (see Bailey 1982; de Bruyn 1997).

Review of Management Practices of Hives with Movable Frames

Steiner commented that beekeepers in the past had a strong instinct for managing bee colonies, for example, in relation to the selection and placement of the hives. He concluded that such innate abilities appear to have been lost (85). He acknowledged that there seems to be benefits for the beekeeper and the bees of framed hives compared to skeps, the basket-like hives, because the honey and wax can be taken without destroying or seriously damaging the bee colony. Nevertheless, Steiner points out that wood is a very different material from plaited straw or similar materials used to make skeps, each attracting different substances from the environment, which in turn affect the bee colonies differently (85–86).

To examine spiritually the impact of hive designs and its associated management practices, it is worth recalling Steiner's observation of a colony as a head without a skull (176–177). He commented that the same processes occur in both a bee colony and the human head, although the human head is protected by the skull, while a bee colony needs to obtain such protection from the external environment. The way bees do this offers great insights for the development of beehives and associated management practices that acknowledge the wisdom and integrity of the bee colony.

Little detailed research is available on the honeybees' natural nest construction. One of the few studies is by T. D. Seeley and R. A. Morse (1976), which examined twenty-one natural bee nests in the U.S. The findings were supplemented with additional information from experiments on bee colony choices of preferred nest sites. This material is the basis for the information presented here. Under natural conditions, bee colonies build their nests in the hollow parts of trees or rock cavities and other areas that provide the colony with protection from the elements and predators. In these cavities bees connect their combs to the walls of the cavity to form a more or less cylindrical enclosed space.

Bee colonies do not seem to express a strong preference for the type of tree in which to build their nests. The study found that oak trees were used most commonly, but this may well have been because they were one of the sturdiest types of trees in the study's location. Bee colonies did show a preference for nesting in living trees, but they would also nest in sturdy dead trees. The shape of a nest was determined largely by the shape of the tree trunk, cylindrical and upright. The average volume of a nest was found to be approximately 1 1/2 cubic feet (45 liters). Bee colonies also preferred a nest cavity of this volume when given a choice during experiments. The majority of nests had their entrance near the bottom end of the tree cavity (a position also preferred when bee colonies were given a choice), with a diameter of around 1 1/4 to 1 1/2 inches (32 to 38 millimeters).

The bee colony coats the interior of its nests with propolis of varying thickness. The colony appears to remove the rotting wood and to fill cracks and crevices with propolis to prevent small predators from hiding inside the nest cavity. The antimicrobial qualities of the propolis stop the growth of microorganisms and prevent further rotting of the tree, thereby increasing its lifespan (Seeley & Morse 1976). This research shows the mutual benefits for both the bee colony and the tree selected by the bee colony for the nest site. The bee colony benefits because the tree provides protection, a kind of skull, or skin. The tree benefits because the bee colony closes the nest cavity, which prevents dirt and moisture from entering the cavity, while the propolis prevents rot.

Under natural conditions, a bee colony stores honey near the top, and pollen is stored below the honey and above the brood. This use of the colony's comb has to do with the winter movement of the bee cluster. During winter, a colony is reduced to a core population of a queen and the workers. This cluster moves upward during winter, consuming honey and pollen in the process. The comb containing drones is located on the periphery of the nest and uses about seventeen per cent of the space. The comb cells for storing the honey may be deeper than the cells of the brood comb (Seeley & Morse 1976).

The comb enables a relatively constant temperature to be maintained in the brood area as air movement is restricted by the enclosed spaces of the comb. The warm air of the nest is kept in at the sides, and draughts and condensation are minimized. The unique way in which under natural conditions the bee colony constructs comb creates a breathing process whereby the carbon dioxide rich air from the bees' respiration drops to the open bottom end of the comb where it is exchanged with fresh air. If the level of carbon monoxide rises, the bees are able to increase the air circulation through the movement of their wings. The sealed sides of the comb prevent cold air from entering (Morse & Hooper 1985; Thür 1946). The way the bee colony creates its internal breathing process provides further evidence of the wisdom of the bee colony's internal workings and therefore reinforces Steiner's observation that a bee colony should be understood as an individual organism.

The effort a bee colony puts into creating its internal environment is essential for the development of the colony's brood, the eggs and larvae. The bee colony needs to maintain a constant temperature of approximately 95° F. (35° C.) and a humidity level of around 40 percent for its brood to develop. Any sudden cooling can result in the brood not developing or not developing properly. The temperature that is required for the successful hatching of brood is higher than the air temperature outside the nest in most beekeeping locations. As a result a bee colony needs to spend a considerable amount of effort to ensure that the right temperature and humidity levels are maintained. This is achieved by the bees converting honey into heat through the muscles when moving their wings. A bee colony is so effective in creating a stable internal environment that it becomes largely independent of the external environment and able to survive in the open if it can avoid predators and weather and other damage (de Bruyn 1997; Morse & Hooper 1985; Thür 1946).

In addition to being characterized by its stable temperature and level of humidity the air inside the nest is also saturated with the scents of the bees' pheromones, propolis, wax, honey, pollen, and enzymes. The particular air quality inside the nest of a bee colony is referred to as nest scent. The combination of warm air and the nest scent creates an environment that suppresses the development of harmful bacteria and minimizes the risk of diseases occurring (Morse & Hooper 1985). Steiner, too, points in his lectures to the presence of substances with antimicrobial qualities in the bees and bee colony.

The way a bee colony selects its nest site and constructs its combs shows that such natural processes and activities are underpinned by logic or wisdom. The importance of these natural processes and activities for the well-being of a bee colony and for maintaining a stable temperature and nest scent has long been recognized (see Thür 1946; Warré 1948). The framed hives that were developed in response to the discovery of "bee space" interfere with a bee colony's natural process of comb building and its activities in maintaining the optimum internal environment needed to raise a brood successfully. Hence, they have been controversial since their introduction.

Critics of moveable-frame hives argue that, when frames are stacked in neat rows in a hive box with straight walls and square corners, it is much more difficult for bees to maintain a stable temperature and nest scent as the air circulates more freely than in a more natural nest environment. Spaces between the frames allow cold air and moisture to enter the hive and easily pass over combs. Because of the increased air circulation, the warm air cools more frequently and forms condensation, a breeding ground for mold.

Critics of these hives also point out that each time a hive is opened the temperature drops and additional effort is required of the colony to raise and maintain the temperature. The situation is exacerbated as warm air rises to the top of the hive and is the first to escape as these hives are opened from the top. The effort needed to maintain a stable temperature and nest scent is further increased by placing empty hive boxes on top of the brood chamber, which also draws heat away from the brood. Such practices mean that a greater number of bees are needed to maintain the internal environment of the hive, and that fewer bees are available to leave the nest to collect pollen and nectar. The need to reestablish the optimum internal environment after each interruption also leads to an increase in honey consumption by the bees. The overall result of practices associated with framed hives is a reduction in the amount of honey available at the end of the season (Thür 1946). Another effect of such management practices is that the frequent interruptions place stress on the colony, which, as we will discuss in more detail later, contributes to an increased susceptibility to pests and diseases.

In their wisdom, bee colonies try to counteract the negative effects of open spaces in framed hives by building "brace combs" with propolis to close the spaces between frames and the gaps between the box and the frames. This makes it difficult for a beekeeper to open a hive, to separate the hive boxes, and to lift the frames. It is common practice to remove these structures, which leads to an ongoing battle between the beekeeper and the colony. Critics see this attempt to fill the open spaces in framed hives as further evidence that such hives are inappropriate for keeping bees. However, most conventional beekeepers do not recognize or accept that the colony is merely trying to rectify an inappropriate situation by creating a more optimum internal hive environment, simply considering such behavior a nuisance (Thür 1946; Warré, 1948).

Thür (1946) observed that the introduction of framed hives also changed the way bees are studied. Similar to earlier observation of the nature of science by Steiner, Thür stated that framed hives allow a better observation of bees inside the colony (albeit in an artificially created environment rather than in their natural state). But this in turn shifted the focus from the colony as an entity to the individual bees within a colony, and from the colony's relationships with the wider environment to the characteristics of individual bees. Consequently, the nature of a bee colony is increasingly explained in terms of the behavior of individual bees. The type of knowledge obtained in this way is reminiscent of Steiner's example referred to earlier: One is trying to explain the operation of a compass by examining only the properties of the compass pointer.

Another consequence of this shift in perception is that it became common practice among beekeepers to open the framed hives to check on the health of the colony. This may be an exaggerated analogy, but in light of the importance of maintaining a stable internal environment for the colony's well-being, opening the hive is like opening a person's chest to see if the heart is still ticking. Clearly, ways are needed to observe the health of a colony without opening the hive.

Critics have concluded that framed hives create significant shortcomings and are harmful to bee colonies. Moreover, framed hives noticeably reduce honey production, a fact that is almost completely unrecognized according to Thür, since most beekeepers no longer understand the bee colony's nature (See Thür 1946; Warré 1948).

Wax Foundations, Queen-excluder Boards, and Smokers

The introduction of framed hives led to the introduction of a number of other practices that further mechanized beekeeping and interfered with the bee colony's natural processes. As mentioned, one difficulty associated with framed hives is that bees do not necessarily follow the intended design of the hive by building their combs only on the frames. They also attach the frames to each other and to the sides of the hive box. This negates many of the perceived advantages of framed hives for beekeepers. This problem was solved when Johannes Mehring

(1815–1878) invented the wax comb foundation in 1857. A wax comb foundation consists of a thin sheet of wax on which the outline of the cell base and walls are pressed. These sheets are strung on the moveable frames to fool bees into constructing their comb based on the predetermined pattern.

The wax foundations also ensure a greater use of the available space on a frame. Under natural conditions, the combs would tend to taper off at the bottom to manage the weight on the upper cells and prevent the comb from collapsing or falling off its support. For beekeepers, the down side of this method is that bees do not fill the frame, resulting in less honey per frame. Wax foundations address this by covering the entire surface of the frame.

Another perceived benefit of wax foundations is their potential to minimize the number of drones. Conventional beekeeping regards drones as useless, despite the fact that drone combs do not seem to have a negative affect on honey production (Seeley & Morse 1976) In fact, the presence of drones may well benefit the colony, as discussed earlier. Despite evidence to the contrary, conventional beekeeping practices aim to decrease the ratio of drones to worker bees to prevent a colony from wasting resources by raising too many drones. One such method is to use the smaller worker cells as the basis for the wax foundation. Bees make comb cells of two sizes (ignoring queen cells for the moment): a smaller one about 1/5 inch (5 millimeters) in diameter for workers, and a larger one almost a 1/4 inch (6 mm.) for drones. Based on the size of the cell, the queen bee decides whether to lay a fertilized worker-bee egg or an unfertilized drone egg in the comb cell.

Since the introduction of wax foundations in the late nineteenth century, manufacturers have increased the cell size from about 0.19 to 0.20 inches (5 to 5.1 mm.), or even 0.21 inches (5.3 mm.), increases of 4.1 to 10.2 percent (Conrad 2007, 142). In terms of human sizes, the height of a person would increase from 5 feet 9 inches (175 cm.) to as much as 6 feet 5 inches (196 cm.), or from about 165 pounds up to about 180 pounds (75 to 82 kg.). It is assumed that such an increase in cell size leads to larger bees, which will collect more pollen and nectar and therefore be more productive. In addition to artificial foundations made from recycled wax, plastic comb foundations have also been developed in recent years. A small coating of wax has to be applied before bees are willing to build honey comb on plastic foundations.

Another issue that beekeepers wanted to resolve is that when bees are left to their own devices, the frames often contain both honey and brood cells. The disadvantage of this is twofold: first, the honey being extracted may contain larvae and eggs, requiring the honey to be filtered; second, some of the brood will be destroyed when the honey is taken. In 1865, Abbé Collin solved this problem with his invention of the queen excluder. The queen excluder consists of a thin screen with parallel gaps large enough to let worker bees through but too small for the queen bee. The screen is placed between the hive boxes containing brood (the brood chamber) and supers intended to store honey and pollen. The queen excluder limits the movement of the queen to the brood chamber, while worker bees maintain access to the other hive boxes.

Because framed hives allow the bee colony to be inspected more easily than those in, say, skeps, they are opened more frequently. The more frequent disruptions to the colony increase the risk to the beekeeper of getting stung. Very early in the history of human interaction with bees, it was discovered that bees dislike smoke. Consequently, smoke has been used to control, pacify, or destroy bees when taking their honey. The pacifying effect of smoke is attributed to disruption of the colony's communication and producing a flight response. This causes the bees to fill themselves with honey in anticipation of leaving the hive. Both aspects reduce the risk of bee stings. The issue of applying smoke more selectively when removing a frame, for example, was solved in 1875, when Moses Quinby introduced the smoker. The smoker is essentially a small bellows that enables the beekeeper to aim smoke into specific areas of the hive. The smoker allows the beekeeper to pacifying the bees more effectively and work more comfortably than was previously possible.

Review: Wax Comb Foundation, Queen Excluder, and Smoker

All three of these inventions—wax-comb foundation, queen excluder, and smoker—interfere with the colony's natural internal activities. Under natural circumstances, the bee colony determines the layout of the comb in the nest, the type of cell to be built, and the size of those cells. The colony determines when to build new cells or break down old ones. Beyond the availability of nectar and pollen, very little is known about the factors that influence these decisions. Bee colonies also determine the ratio of drones to workers in a colony.

These interventions occurred despite an incomplete understanding of the role of drones. Nor is there a good understanding of the long-term effects on the well-being of the colony from the use of fixed-size cells and wax foundations made of recycled wax for brood cells. Because wax foundations are made of used honeycombs, they may contain traces of chemical pollutants and other impurities that become part of the new cells for raising the brood and for storing honey. Evidence indicates that bees are not very fond of wax foundations in the hive, and if they are not supplied at the right time, bees break them down as they do with other old wax structures (Bailey 1982, 29).

There is some indication that the increased cell size of wax foundations assists the Varroa destructor (an external parasitic mite) in its development, and that by reducing the cell size, a colony is able to minimize varroatosis, the disease caused by such an infestation. A possible reason for this is that the increase in the bees' size results in a decline in their vitality. Furthermore, Varroa appears to be less destructive to smaller bees, such as the Asian honeybee (*Apis cerana*), its natural host (Conrad 2007, 140–142).

For beekeepers who wish bees to build their own natural combs, it is important to recognize the evidence suggesting that bees raised for generations on wax foundations lose their ability to build combs naturally and may require a gradual transition from wax foundation to natural comb construction (Conrad 2007; Hauk 2008).

The queen excluder is advantageous for beekeepers who wish to have frames filled only with honey and avoid the destruction of eggs and brood during the honey harvest. There may also be a consumer preference for honey extracted from "new" cells instead of those that previously contained eggs and larvae. Such a preference, of course, ignores the fact that the wax used in foundations is old and recycled.

In their natural state, all the bees in a colony are able to move freely throughout their nest and, as mentioned, do so in a way that follows an annual rhythm. At the end of winter, the bees can be found near the top of the nest, as they have eaten the honey stored during summer. Once spring starts, the colony moves downward, filling the old comb first with brood and next with pollen and honey, moving their way down through the year. During winter, the colony slowly moves up again as it consumes its stores of honey (Thür 1946). Queen excluders interfere with this annual rhythm. They also prevent the queen's ability to move close to the colony's food stores.

The queen-bee excluder also forces the bee colony to engage in another practice contrary to its nature. Under natural conditions, a bee colony builds new cells in which the queen bee lays her eggs, using the older cells to store honey and pollen. This strategy minimizes the negative effects on the newly developing bees from contamination or impurities left over from earlier hatchings or from honey or pollen that may carry harmful spores or be otherwise contaminated. Under natural conditions, bees avoid storing honey in new cells. The queen excluder, however, forces the unnatural behavior of building new cells above the brood to store honey and forces the queen to lay her eggs in used comb cells.

The use of a smoker greatly improves the beekeepers ability to work with a bee colony. However, smoke is stressful for a colony. Its negative effects are further influenced by the type of material being burned, since smoke may contain pollutants that can poison bees and damage the combs. Some materials have been found to give off less harmful smoke than others (see Conrad 2007). Too much smoke or smoke that is too hot can also damage the bees and the comb.

Alternatives for pacifying bees include spraying a mist of water over the bees and the comb, which diverts the bees' attention away from the disturbance and toward drying themselves. Another method is drumming, a traditional method of moving bees from one part of a hive to another. Drumming involves quick, regular tapping with both hands or with sticks on the part of the hive from which the bees are to be moved. Regular tapping creates a hypnotic effect, and bees are said to move away from the drumming without agitation (see Crane 1999; Warré 1948).

A Way Forward: Alternative Hive Design and Hive Management Practices

Framed hives have led to the introduction of a range of practices that interfere with or prevent the natural functioning of a bee colony and, as a result, were controversial when introduced. To avoid such problems requires the development of different management practices and hive designs. The bee colonies' selection of nest sites and the way they construct combs and manage their environment within the nests clearly show their wisdom. Acknowledging and respecting that wisdom in the design and construction of hives acknowledges and supports the bee colony's integrity and its role of revitalizing plants, pollination, collecting nectar, and producing honey. Steiner's research and findings, as well as those of others with a similar sensitivity to the nature of the bees, show that one central principle to guide the development of more harmonious designs is the importance of minimizing the frequency and degree of interference with a bee colony.

It is interesting to note the current requirements of the International Demeter Standards for Beekeeping and Hive Products (2008) with respect to hives and comb. These require that beehives be built entirely of natural materials such as wood, straw, or clay. The only exceptions are for the fixings, roof covering, and wire mesh. Under this standard, the beehive interior must be treated with beeswax and propolis obtained from Demeter (biodynamic) beekeepers, while the exterior must use natural, non-synthetic, and ecologically safe wood preservatives.

The Demeter standards also state that all combs should be constructed naturally without the help of wax foundation midribs (the surfaces on which the cells are constructed). However, it does allow both fixed or movable frames and strips of beeswax foundation to guide comb building. Only in the supers where honey and pollen are stored may wax midribs be used, but the standard states it is nonetheless desirable to avoid their use.

By permitting the use of moveable frames and wax strips, this standard still permits a high level of interference associated with managing framed hives.

With respect to the hive's brood area, the Demeter Standards state that both comb and brood area must be able to grow naturally as the bee colony develops. This means that the dimensions of the brood chamber and frame size must be such that the brood area can expand without being obstructed by wood from the frames. Barriers that separate the brood are not permitted.

In practical terms, this means that the brood chamber of a hive should accommodate a brood chamber of the size that occurs naturally, without internal barriers. Without human interference, the size of the brood comb will expand from about 10 to 16 inches

(25 to 40 cm.) in diameter (see Warré 1948; Weiler 2005). The size of the brood depends on various factors: the time of year, the health of the colony, and the breed or race of the bee. As discussed, under natural conditions the brood "moves" throughout the year, either from top to bottom in vertical hives or from front to back in horizontal hives. To meet the brood's requirement, Weiler (2005) suggests frames of 13 3/4 inches square (350 mm.).

An Example of Alternative Vertical Hive Designs: The Warré "People's Hive"

Fortunately, alternatives to framed hives have been developed parallel to moveable-frame hives. One well-known hive design is by Abbé Warré (1948). It is based on his observations of the natural behavior of bee colonies and by testing a range of designs in existence at the time.[1] His findings and the reasons for developing his particular hive are detailed in his book, *Beekeeping for All* (translated by David Heaf from *L'apiculture pour tous*). Abbé Warré's design is based on the importance of maintaining the nest scent in the hive.

The Warré hive goes a long way toward addressing the negative effects of conventional beekeeping practices and hive designs, as it reduces disruptions to the colony to only two times a year: adding hive boxes in spring and harvesting honey at the end of the season. Compare this to once or twice each month for inspections during the foraging season as is typically advocated in conventional beekeeping. Abbé Warré's design allows bees free access to every part of the hive, and it allows the colony to build natural combs for rearing brood and for storing honey and pollen. The design does not manipulate the ratio of workers to drones, nor does it interfere with the colony's natural reproductive processes. Details on the design and management practices associated with the Warré hive are discussed in the addendum to this chapter.

---

1. Abbé Warré hive design is based on thirty years of studying 350 hives designs, including skeps and moveable-frame hives such as the Dadant.

Skeps and Warré hives.

Nonetheless, this design has two aspects that present potential problems: 1) escape of nest scent and a drop in internal hive temperature during the honey harvest, and 2) the fact that the brood is spread through two boxes with fixed bars running through the middle of the brood. The latter contravenes the International Demeter Standards for Beekeeping and Hive Products (2008). The loss of temperature and nest scent around the brood during the honey harvest can be reduced by placing a division board between the top hive boxes that are to be harvested and the three boxes that are left for the colony for winter. To enable bees caught between the hive roof and the division board to get away, bee escapes can be places between the top hive box and the roof.

An Example of Alternative Horizontal Hive Designs: Top-bar Hives

Top-bar hives are alternatives to the upright hive design (see Chandler 2006). They evolved from cylindrical hives and can be constructed easily in a size that allows the brood access to the

A top-bar hive and clay-cylinder hives.

comb and food stores during winter. Top-bar hives, too, enable bees to build natural comb and allow full access to every part of the hive. However, top-bar hives usually open from above, enabling nest scent to escape and the internal temperature of the hive to drop. One possible way to reduce this problem is to modify the design so that the top bars can be removed from the front or the back, as is done with traditional cylindrical hives.

Beekeepers continue to experiment with hive designs to find the best balance between the needs of the bee colony and the beekeeper. For example, Gunther Hauk (2008) of Spikenard Farm,[2] draws on Rudolf Steiner's observation of the ways that shapes, forms, and materials in nature affect various hive designs, including round and top-bar hives.

The appendix to this volume outlines criteria for consideration when developing alternative hive designs that are in

---

2. Spikenard Farm and Apiary in Illinois is a not-for-profit organization founded by Gunther Hauk and Vivian Struve-Hauk. A core group works in the fields and pastures, as well as in a market garden and greenhouse, and maintains honeybees. The farm offers apprenticeships and internships in biodynamic beekeeping.

keeping with the research findings of Rudolf Steiner and others who acknowledge the wisdom of bees. Incorporation of such research in hive design is relatively recent, and research into the effects of the different features is still needed. When designing hives, it is important to be aware of laws and zoning regulations related to beekeeping.[3] Ideally, beekeepers would benefit from discussing and exploring the characteristics of different hives with others in similar climatic and local conditions. In theory, a bee colony can survive for a long time if it is able to renew is comb body regularly. One case mentions a century-old bee colony (Baker 1948, 14).

The importance of interfering as little as possible with hives also means that beekeepers need to develop observation techniques that allow them to assess the well-being of a colony without opening the hive. Just as farmers are able to assess the health of their animals according to their behavior, without cutting open or opening up the organism, beekeepers need to develop the ability to observe colony behavior from outside the hive.[4]

Discussion

Rudolf Steiner's research and similar research by others show the wisdom behind the activities of bee colonies, both in relation to the external environment in the selection of nest sites and in relation to creating an internal environment through the particular construction of the comb. When the natural process of a bee colony is compared to conventional beekeeping practices, it quickly becomes clear that such practices give little recognition to the wisdom behind the natural behavior of the bee colony.

The use of framed hives and their associated management practices create an internal environment that requires a bee

---

3. In Australia, for example, it currently not permitted to keep bees in hives other than hives with moveable frames or top bars.

4. In addition to the examining nature spiritually, as outlined by Steiner, books such as those written by Warré (1948) and Weiler (2006) contain information on such diagnostic skills.

colony to expend great effort to maintain the proper temperature and nest scent for brood development. The situation is made worse by the practice of placing additional hive boxes on top of the brood chamber (causing the warm air and nest scent to flow away from the brood into the empty top box) and by opening the hive for regular inspections. This also allows the temperature to drop and the nest scent to escape. To compensate for a temperature drop and loss of nest scent, a bee colony must expend extra effort to reheat the hive, which decreases its food supply because fewer bees are available to collect pollen and nectar. The loss of heat also increases the consumption of winter stores and the need for more honey or artificial nutrition for the bees.

Steiner mentioned that in earlier times beekeepers had an instinctive knowledge for beekeeping practices, much of which has disappeared. At the time of Steiner's lectures, the impact of unnatural influences such as radiation, electromagnetic waves, and chemicals were far less than they are today. Though little conclusive research is available on their effects on the well-being of bee colonies, it is a good idea to bear in mind the potential impact of these factors when selecting the location and aspects of bee colonies.

The International Demeter Standards for Beekeeping and Hive Products (2008) promotes the management of bees in accordance with the findings of Steiner's research and addresses many of the shortcomings associated with conventional beekeeping. The task of designing and developing hives that fully respect the bee colony's wisdom is not an easy task, and even hive designs that go a long way toward that goal, such as the Warré hive, still have a number of shortcomings in that respect. The challenge is to meet both the needs of the bees and those of the beekeeper. However, because throughout much of the recent history of beekeeping the needs of the beekeeper have taken precedence, the decline of honeybee populations around the world provides a strong incentive for putting the needs of

HIVE DESIGNS AND ASSOCIATED MANAGEMENT PRACTICES

HIVE FEATURES & MANAGEMENT PRACTICES	HIVES WITH MOVEABLE FRAMES	HIVE DESIGN BASED ON MINIMAL INTERFERENCE WITH THE NATURAL PROCESSES OF THE BEE COLONIES
MATERIALS	Mostly wood; recently other materials such as plastic and metal have been suggested. Paints and other preservatives are used to protect the hives, and metal elements are used in construction.	Wood or wicker, covered with clay or mud. If small amounts of metal are used, perhaps copper may be the most appropriate metal with bees, owing to its association with Venus.
SHAPE	Oblong	Square or round, positioned horizontally (tree trunk) or vertically (rock cavity).
MANAGEMENT	Opened several times a year for inspection.	Minimal opening; every effort should be made to maintain temperature and nest scent.
FRAMES	Movable.	Fixed top bar or none.
LOCATION	Sheltered from the elements.	Considers environment and cosmic influences.
NEST SCENT	Largely ignored and not considered in practices.	Aimed at maintaining temperature and nest scent.
WAX FOUNDATION	Used to manipulate colony's composition (worker bees and drones) and bee development (size). Made from recycled wax.	None; bees must build their own comb.
QUEEN EXCLUDER	Used to restrict movement of the queen to the honey stores.	None, bees are able to move across the comb.
SMOKER	Used to pacify and manipulate bees.	Usually none; use drumming and water spray instead. Manipulation to empty the honey boxes achieved through drumming; bee escapes.

bee colonies first when developing beekeeping practices and designing beehives.

The adoption of hives that minimize interfering with the colony requires development of observation techniques and diagnostic skill to assess the health of the colony from the behavior of the bees outside the hive.

The table on the previous page summarizes conventional beekeeping practices associated with framed hives and beekeeping practices based on minimizing interference with the bee colony as informed by spiritual research.

ADDENDUM: ORIGIN OF THE PEOPLE'S HIVE (THE WARRÉ HIVE)

The Warré hive generally consists of three boxes of equal dimensions and a roof and bottom board. The hive's brood box consists of two boxes with an internal dimension of 12 x 12 x 8 1/2 inches (30 x 30 x 21 cm.), creating a volume of approximately 1.25 cubic feet (36 liters), with the third box of the same size to store honey and pollen. Each box has eight fixed top bars on which the bees build their combs. The square shape of the brood boxes was chosen by Warré as the shape closest to the natural cylindrical nest cavity found in trees. A cylinder enables a more even distribution of heat within a space than does the oblong shape of conventional hives. The bees therefore expend less effort to stabilize the hive temperature. Warré selected square hive boxes as a compromise, as these are easier and cheaper to construct and maintain. Inspections occur by observing the colony's activity from outside the hive. The inclusion of glass walls in the hive boxes enable the beekeeper to observe what is taking place inside the hive. The glass wall is especially useful for inspecting if the hive box contains the brood or honey and pollen just prior to its removal at the end of the season.

Warré found that the volume created by the dimensions of his two hive boxes (12 x 12 x 16 1/2 inches, or 30 x 30 x 42 mm.) enables the winter cluster to move upward during winter and consume the stores above, which is their natural preference. His reason for deciding on two boxes rather than one is to make harvesting the honey easier. A larger hive box is more likely to contain both honey and brood at the bottom of the comb. If the larger box is split into two boxes, the box with the comb containing the brood can be left with the colony. The square shape also enables access for bees to upper stores during winter. Bees find it difficult to access honey stores to the side of the cluster, which is required in hives with the same volume but with shallower dimensions. A square shape also enables the top bars to be placed cold-way (at right angles to the hive entrance) or warm-way (parallel to the entry wall). The former enables air to move more freely through the hive and is useful during summer, whereas the latter prevents air circulation and is better during colder seasons. Conventional hives frames are generally placed cold-way. Warré acknowledged one shortcoming of his design—that, for over-wintering, the brood chamber consists of two parts, with top bars in the middle, and thus leaving open space (Warré 1948, 51).

Management practices associated with the Warré design manipulate the hive only twice a year: once to extract honey and once to add empty hive boxes. Generally, two additional hive boxes are placed below the three boxes. By placing them below the existing boxes, the nest scent is left relatively undisturbed, because the brood box is not opened from the top. Throughout the year, the colony moves down, building new comb for the brood and for storing honey and pollen in the older combs. At the end of the season, the top hive boxes will contain mostly honey and pollen and can be removed without taking combs containing brood. At least one box of honey and pollen is left to feed the colony during the winter. With the exception of the two annual disturbances, the hives are left alone.

6

ARTIFICIAL BREEDING

*This whole procedure must not be carried out in a way
that is too rational and businesslike...we'll see that what
proves to be an extraordinarily favorable measure upon
which something is based today may seem good, but that
a century from now all breeding of bees will cease if only
artificially produced bees were used.*

—RUDOLF STEINER (178)

At the time of Rudolf Steiner's lecture of November 10, 1923,
the practice of artificially breeding and the rearing of queen
bees had been in existence for about twelve to fifteen years. Mr.
Müller commented,

I hold this method in high regard for the most part. If
you leave the colony to its own devices and don't tend it
carefully, the whole colony might deteriorate. The bad
qualities will come through more and more, and whatever
was good before is lost. (177)

Since that lecture, the narrow focus on satisfying economic
objectives has continued and has resulted in even more dra-
matic interference and manipulation in the bee colony's natural
reproductive processes. This has reached such a level that today

beekeepers typically interfere with virtually every stage of the natural reproductive cycle of bees, leaving little of the natural reproductive process in conventional beekeeping.

Bee Breeding

As mentioned, the reproduction of a bee colony consists of two processes: new bee colonies are "born" through swarming, and individual bees are raised to insure the colony's continued existence. Swarming involves the established queen, worker bees, and sometimes young queens leaving the hive to establish a new colony elsewhere. In times of abundant nectar and a strong colony, workers begin to build a number of swarm cells in which the queen will lay her eggs. Just before or shortly after one of the swarm cells hatches, the queen leaves the colony. Depending on the health of the established queen she is either quickly replaced by one of the new queens that accompanied her or she remains as part of the new colony. In the old colony, the first move of the new queen upon hatching is to kill the larvae in the other queen cells, as well as the other newly hatched queens, until only one remains.

Colonies can also replace a queen bee without swarming. This is called "supersedure" and generally occurs when the established queen bee is no longer healthy or able to lay fertile eggs, thus placing the colony's survival at risk. In such instances, the colony raises a new queen bee to supersede the established one, enabling the colony's continued existence. If the existing queen bee happens to be killed accidentally, the colony can raise a new queen from existing worker eggs by expanding their cells and providing additional nutrition to speed development. Usually such a queen is not as strong as one from an egg laid and hatched in a queen cell, and she is an interim measure and often superseded once a queen bee can be raised in the normal way.

Historically, the ability to swarm was considered a desirable trait, as it enables the number of colonies to increase. This was especially true when harvesting honey involved the colony's destruction. Since bee colonies are no longer destroyed to harvest honey, beekeepers have come to regard swarming as less desirable, as it reduces the strength of the colony by dividing itself into two or more new colonies. Both the established and new colonies need to spend time and energy building up their strength at the expense of gathering nectar and producing honey.

Another perceived disadvantage of the bee colony's natural reproduction process is that, because mating occurs high in the air and in the open, the beekeeper has little control over the drones with which a queen bee chooses to mate. Hence a beekeeper has little influence on the characteristics of the future bee colony.

Artificial Queen Breeding and Queen Rearing

At the time of Steiner's lectures, several different practices existed for breeding and rearing queen bees. Generally speaking, after the original queen has been removed and killed, one would manipulate the mating process and introduce the newly mated queen bee into hive.

Usually, the artificial queen-raising process today is as follows: Artificial queen cells (queen cups) replace the queen cells normally constructed by the bee colony. They are usually made of plastic, with a depth and diameter of about 2/3 inch (10 mm.) and 4/100 inch (1 mm.) thick. Next, the artificial queen cells are connected to a cell bar, and some diluted royal jelly is placed in each cell along with a day-old larva transferred (or "grafted") from a worker cell. The grafted cells connected to frame bars are placed into a starter colony without a queen for twenty-four hours, while the nurse bees finish building the artificial queen cells and feed the larvae.

After twenty-four hours, the artificial queen cells are placed into a finishing colony, in which the cells are fed for about three and a half days before they are capped by the nurse bees. These colonies, too, have usually had their queen bee removed and killed so that bees will make a greater effort to look after the new queen cells. During this period, the bee colonies that will receive the grafted cells are fed sugar syrup to ensure that the bees are able to produce enough royal jelly to feed the larvae.

The queen cells can be left in the finishing colony until a day before they are due to emerge. Shortly before this, the queen cells are placed into small colonies called mating nuclei, where they either undertake a mating flight or are artificially inseminated, generally three to five days after they have emerged. The queen bees either remain in the nuclei colonies, which are allowed to develop, or they may be introduced into other colonies.

The discovery that queen bees can be raised from worker eggs and transplanted into queen-less colonies meant that fertilized queen bees could be transported over long distances. This led to the present-day industry practice of artificially replacing a queen bee every one or two years, irrespective of her health or productivity or the fact that the queen's natural productive span is more than double this period. As a result of such interventions, virtually all queen bees originate from eggs intended to become worker bees, with most beekeepers purchasing their queens from breeders.

Artificial Insemination

In addition to artificially rearing queen bees, the narrow focus on genetics as the determining factor of a bee colony's characteristics has also led to the development of practices to manipulate the natural mating processes by determining which drones will fertilize the queen bee. Some of the characteristics seen as desirable by conventional beekeepers are high productivity, a minimum use of propolis, reduced swarming, and, more recently, pest and disease resistance.

Because a queen bee is responsible for half the genetic makeup of a colony, with the other half coming from the drones with which she mates, beekeepers prefer to select those drones.

These come from colonies that have been manipulated to create drones by providing them with drone combs from other colonies or by using wax foundations about thirty-five to forty days before they are needed, to coincide as closely as possible with the emergence of the artificially raised queens.

Once semen has been extracted from the selected drones, the queen is inseminated. Prior to insemination, the queen bees are anesthetized with carbon monoxide, which facilitates the deposit of semen by syringe into the queen bee's reproductive organs. Without being anesthetized, this process is more difficult, as queen bees are naturally less cooperative (see (Laidlaw & Page 1997).

The newly inseminated queen bees are transplanted into a colony from which the established queen has been removed and killed to facilitate the acceptance of the artificially inseminated queen. If the queen bee is accepted, the adoptive colony starts to show the characteristics of the new queen and the drones with which she mated after about twenty-one days, the time it takes for the first new worker bees to develop.

Review of Artificial Queen Breeding Practices

Apart from any ethical concerns about the treatment of animals, artificial bee breeding practices have little regard for the intricate interrelationships within a bee colony. Steiner saw artificial breeding inappropriate: "Certain forces that have operated organically in the beehive until now will become mechanized" (21). By introducing an artificially fertilized queen bee into a colony, the strong connection that nature established between the queen bee and the worker bees and drones is disturbed. Steiner states that it is impossible for a bee colony to establish the kind of intimate relationship that occurs naturally with a queen bee that is purchased elsewhere (21).

Conventional beekeepers see the introduction of a queen bee into a colony as somewhat like introducing a dairy cow into an established herd. Steiner's research showed, however, that the relationships among bees in a bee colony are much more intricate, and that a queen bee can be compared more appropriately to a human or animal organ. In fact, for the bee colony the effects of introducing an artificially bred queen bee is more like an organ transplant.

In addition to disrupting the intimate relationships among the bees in a colony, rearing queen bees artificially can also undermine the long-term viability of the honeybee, as otherwise nonviable queen bees will raise colonies. Under natural conditions, a bee colony builds from ten to fifteen queen cells and, after the existing queen has swarmed, the queen that hatches first either forms a second swarm or, if the colony is not strong enough, kills all the other queens. Thus, only the strongest queens survive. Artificial queen breeding interferes with this process of natural selection, with unknown long-term negative effects.

Current breeding practices have led most beekeepers to purchase their queens from breeders. These queen bees are bred in environments alien to the colony they will populate and may not be appropriate for the local conditions. Moreover, this practice has reduced the genetic diversity among honeybees. A small number of breeders supply a large number of commercial breeders, and most commercial beekeepers keep only two of the many subspecies: the Italian bee (*Apis mellifera ligustica*) and the Carniolan honey bee (*Apis mellifera carnica*). The result is that many original European bee subspecies that had adapted to local conditions have become rare and even extinct. (See the addendum on bee species at the end of this chapter.)

Steiner's research identified the possibility of changing worker bees into queen bees by manipulating the period for their development and nutrition. Müller, who embraced modern beekeeping practices, referred to a queen bee developed in this way as a "pseudo queen," saying that it is the result of a hive that's diseased and that it will never be like one raised under normal conditions

(114). Steiner also pointed out that, when a bee colony changes a worker egg and larva into an egg-laying queen, this should be regarded as a type of illness of not just of the queen bee, but also of the whole colony. He compares the phenomena of raising queen bees from worker eggs to force-feeding a goose to develop its liver: "The powers of the liver are developed to an abnormal degree and the entire organism becomes sick. If you make a worker into a queen, then this queen is in reality an overly healthy worker, but the entire beehive should be considered ill" (115).

At the time, Steiner did not foresee that what he considered diseased colonies would become predominant throughout the Western world. He stated, "All of this goes to show you how mutable such an animal is. But this doesn't have much of an influence on the practical aspects of beekeeping" (114). Hence, Steiner and beekeepers of his time such as Müller would consider most bee colonies of today to be sick.

A Way Forward in Bee Breeding

Steiner acknowledged that, despite the apparent benefits of breeding bees artificially, it should occur in harmony with the natural processes of the bees' reproductive cycle and with minimal interference in the colony. The conventional breeding practices discussed here show little consideration for the bees' existing natural reproductive processes. They consider the queen bee to be an individual organism rather than an integral part of the bee colony's interrelationships. The Bee colony is considered a kind of machine with the queen as a mere component. As with all machines, most beekeepers assume that, by "improving" one component, the whole machine is improved.

Steiner's research shows that living organisms are comprised of both spiritual and physical dimensions and operate in accordance with a certain underlying wisdom. As such, organisms such as bee colonies are greater than the sum of their components, and cannot be understood by studying their individual components in isolation. Furthermore, those parts cannot be manipulated successfully without understanding the effects of this on the organism's complex interrelationships. Understanding what defines an organism is reflected in biodynamic animal breeding. This work has been undertaken predominantly with dairy cattle, but also applies to other forms of animal and plant breeding.[1]

At the basis of the spiritual-scientific approach to animal and plant breeding is an acknowledgment of the unity and integrity of a living organism's spiritual, physical, and developmental dimensions. It recognizes that any change in one of the organism's characteristics will affect the interrelationships among all other characteristics. Breeding objectives and strategies to breed selectively for particular characteristics need to consider the impact these will have on the organism's physical characteristics, on its environment, and on the species' unique evolutionary development, or *typus*.

Conventional plant and animal breeding regards an organism's genes as the main determinant of its characteristics and assumes the organism has only a passive role in their physical expression. This has led to a narrow focus on understanding and manipulating an organism's genetic makeup to achieve the expression of certain characteristics. The role of the environment in how the genetic composition of an organism will express itself and its evolutionary process is thus largely ignored. By contrast, for spiritual science an organism's genetic makeup merely provides the multitude of physical possibilities within which the organism can express itself. Organisms and species actively create themselves from these possibilities by interacting with the environment; they have an active role in

---

1. For a discussion of this research, see T. Baars, A. Spengler and J. Spranger, 2003, "Is There Something Like Bio-dynamic Breeding?"; paper presented at the workshop, "Initiatives for Animal Breeding in Organic Farming in Europe and Evaluation for Future Strategies," Oct. 17–18, 2003 (The Louis Bolk Institute, Driebergen, the Netherlands).

their own creation and evolutionary development. By focusing on interactions with the environment, the biodynamic approach to animal and plant breeding highlights the central role of the environment in how an organism expresses its characteristics (Spengler 1997).

The more the environment reflects the organism's optimum living conditions, the better an organism's intrinsic physical and spiritual potential can be expressed. Steiner explained the various subspecies of bees as expressions of the differences in development of the *typus* in response to the different regions in which they developed (90).

To adopt a spiritual, biodynamic approach to bee breeding means enabling a bee colony to express all of its intrinsic qualities by optimizing its environment. The interaction between a bee colony and its environment can be influenced by beekeeping practices such as hive design, for example. Bee breeding should therefore be closely integrated with the other beekeeping practices of a beekeeper and not occur in isolation or external to the local environment, as is currently the case with queen bees sent by mail all over the globe. Ensuring the existence of an optimal environment for bee colonies to express their best characteristics involves bee breeding and beekeeping practices that are in harmony with the bees' nature.

Steiner discussed the importance of economic conditions during a particular period on the adoption and development of beekeeping practices. This applies also to bee breeding. It is important to recognize that the environmental factors to consider in developing bee breeding objectives and strategies include social and economic ones. Improvements in the characteristics of a bee colony involve a combination of the colony's qualities in the context of current economic and social conditions, which may emphasize certain characteristics as beneficial at one time and less so at another time. With respect to swarming, for example, in order to increase honey production, beekeepers have focused on breeding bees that are less likely to swarm. The additional honey produced by preventing swarming is worthwhile economically, however, only if the cost of breeding and purchasing bees is less than the additional returns from increased honey sales. If, say, at some time in the future the economic advantage disappears or it is no longer possible to purchase and transport queen bees by mail, swarming will once again be regarded favorably.

Another aspect to consider is that breeding for certain characteristics reduces the diversity of the species. To breed bees with certain qualities that fit the economic objectives of a particular period may lead to the loss of other qualities that may offer advantages under different economic conditions. Enabling bees to cope with changing environmental, economic, and social conditions, increases their chances for long-term survival, and requires diversity both genetically and environmentally.

In adopting a spiritual approach to bee breeding, we also need to understand how a change in one characteristic of the colony might affect the other characteristics. To use swarming again as an example, reducing the ability to swarm causes bees to build larger colonies and to put their efforts into producing more honey. However, bigger colonies also may increase the presence of pests and diseases, which in turn prevent the colony from gathering nectar and producing honey. Thus, breeding colonies that swarm less is not beneficial unless it is accompanied by an increase in the colony's cleaning behavior and its pest and disease resistance.

Natural reproduction

The only way that the number of bee colonies can be increased without manipulating their natural reproductive process is through swarming. This is reflected in the International Demeter Standards for Beekeeping and Hive Products (2008), which considers artificial queen breeding practices to be inappropriate and prohibits breeding by grafting, artificial and instrumental insemination, as well as the use of genetically modified bees. However, it does allow the creation of an artificial swarm with the colony's existing queen by preempting swarming, and it allows for increases in the number of bee colonies by dividing the remainder of the hive into artificial swarms or scions. The standard also considers the selective breeding of bees through the production of queen cells as part of the swarming instinct, and the replacement of an old queen through the swarming process is therefore permitted for breeding purposes.

These practices that interfere with the bee colony's natural reproduction process are probably included in the standard as a response to the fact that swarming is often regarded as a nuisance by beekeepers, since it is not always possible to capture a swarm. For beekeepers who live away from their hives, swarming is perceived as even more troublesome. In order to maintain bees in situations where swarming is inappropriate, it has been suggested, in keeping with the International Demeter Standards (2008), that the colony be split around the time it is producing queen cells, the period when it is preparing to swarm naturally (Hauk 2008).

However, the strategy of splitting a hive to prevent swarming interferes with the colony's natural reproductive process and prevents the swarm from being exposed to the physical and spiritual forces from the cosmos and the Earth. If the location where bees are kept means that the colony's natural reproduction process cannot take place, this may indicate that the location is not suited for beekeeping.

In light of the importance of swarming for a bee colony, perhaps beekeepers who are unable to tend their colonies during this period of the year should reconsider their involvement with beekeeping. The period when animals give birth is a challenging time for most people who keep livestock, and there are relatively few animals that can be left entirely to their own devices during this time. Both dairy and sheep farmers must

A bee colony being born—
a swarm of bees in a
stand of bamboo.

keep a close eye on their animals during this time of the year. Beekeepers, too, should accept that the birth of a new colony requires a watchful eye. One advantage beekeepers have is that bee colonies do not give birth (swarm) at night, and unlike dairy farmers beekeepers are assured of a good night's sleep. In countries such as Australia, where the European honeybee is an exotic species, additional care is needed to capture swarms, as the impact of exotic bees on native plant and insect species is not well understood. Feral swarms are often regarded by the general population with the disdain often accorded feral cats and dogs.

It is worth exploring alternative bee breeding practices by considering strategies that influence a colony's swarming behavior. One such strategy is based on the fact that bee colonies are more likely to swarm when they are outgrowing their existing nest. To promote swarming, one can limit the size of the hive by adding fewer additional hive boxes to the most productive colonies, thereby encouraging them to swarm. When swarming occurs, these can then be hived. The colonies that are less productive are less likely to swarm, since they have enough available space to store their pollen and honey.

Swarming can also be affected to some extent by factors less easily influenced, such as the flow of nectar. To prevent the loss of a swarm, the beekeeper can encourage swarms to settle in bait hives. Japanese beekeepers, for example, traditionally

Hiving a bee swarm (from Abbé Warré, Beekeeping for All).

place plant varieties that attract the bees near the entrance of an empty hive. Another strategy consists of placing empty hives near a colony, or as happened in earlier times, preparing tree cavities in areas where tree beekeeping was practiced. Other examples that might help in hiving swarms is to plant trees and plants such as bamboo near the apiary site, from which a swarm is relatively easy to collect. Some beekeepers also use a mirror to shine a beam of light into the swarm or spray a mist of water onto the hive to make it settle.

Strategies for encouraging swarms to settle are quite important for beekeeping practices that respect the nature of bees. The physical and spiritual significance of a swarm leaving its nest and exposing itself to the physical and cosmic influences out in the open should not be underestimated in the formation of a new colony.

Natural swarming as a strategy to increase the number of bee colonies has the advantage of minimizing the need to purchase queen bees and promotes breeding within the local environment. In his lectures on agriculture, Steiner stated that a farm should aim to be a self-contained entity that does not depend on external inputs. In beekeeping, a similar view can be taken, whereby the beekeeper supports bees in their natural reproductive processes and does not rely on outside sources through the purchase of queen bees.

As in the case of beehive designs that meet the needs of the bee colony, adopting bee breeding practices that are in harmony with the natural reproductive process of bees also requires the development of a range of observational and diagnostic skills to manage the bee colony's swarming behavior.

Discussion

Nowhere is the interference and manipulation of the natural processes of a bee colony more evident than in the current queen bee breeding and rearing practices. They furthermore bring out sharply the contrast by which conventional science approaches nature and the study of nature from a spiritual perspective. The effect the different scientific approaches have is summarized in the table on the next page. Steiner acknowledged that, despite the seeming benefits to artificially breeding bees, it should be done in harmony with the natural processes of the bees' reproductive cycle and with minimal interference with the colony.

Conventional bee breeding practices show little regard for the wisdom with which bees' have developed their natural reproductive processes. The focus of conventional bee breeding is predominantly on the manipulation of the queen bee and its genetic makeup rather than on working with the holistic relationships between the queen bee as part of a bee colony and a bee colony as a component of a local environment.

The conventional approach to bee breeding regards a colony more as a machine than as a living organism and ignores the potential impact that a change in one part of this complex living

COMPARISON OF ARTIFICIAL BEE BREEDING
AND NATURAL BEE BREEDING

ARTIFICIAL BEE BREEDING	NATURAL BEE BREEDING
Artificial breeding and rearing practices of the early twenty-first century are used.	The bee colony's natural reproductive process is allowed.
The beekeeper determines which characteristics are to be reproduced.	The colony determines whether to reproduce.
The beekeeper determines the characteristics of drones for artificial insemination.	The queen determines whether to mate with a number of available drones from various colonies.
The beekeeper prevents swarming.	Swarming is allowed.
The beekeeper determines timing and number of new queen cells to be built.	The colony determines timing and quantity of new queen cells to be built.
The beekeeper forces eggs intended to become workers to become queens instead.	The colony (queen) determines which eggs will be laid in queen cells.
The beekeeper prepares the queen cells out of plastic.	The queen cells are made of new wax in the shape and size determined by the colony.
The queen larvae are fed artificially prepared nutrients (royal jelly).	Quantity and quality of nutrition (royal jelly) is determined by the colony.
The beekeeper determines which queens will head up new colonies.	The colony determines which queens will survive to head up the new colonies or the existing colony.

relationship will have on other parts and on the relationship as a whole.

Steiner's research shows that an organism consists of both a physical and a spiritual dimension. The characteristics expressed by the organism are the outcome of the organism's response to the external environment. The better the match between the external environment and the organism, the better the organism is able to show all of its qualities. Beekeepers play a significant role in the creation of a bee colony's environment in their choice of beekeeping practices.

The adoption of a spiritual approach to bee breeding also recognizes that economic, social, and environmental conditions change, and that breeding strategies should therefore aim to enable a species to adjust effectively to these changes by maintaining diversity of both the characteristics of bee colonies themselves and the environments in which they occur.

The current International Demeter Standards for Beekeeping and Hive Products (2008) acknowledges the holistic nature of a bee colony and the need for swarming to occur as part of a bee colony's natural reproductive process. However, it still accepts a degree of interference with this process, which prevents the bee colony from fully expressing itself.

To adopt bee breeding practices in accordance with the bee colony's nature minimizes interference with the hive and requires the development of observation techniques and diagnostic skills that enable the beekeeper to manage this process. These skills are no longer part of conventional beekeeping.

ADDENDUM:
THE MOST COMMON HONEYBEE RACES OR BREEDS

There are a number of honeybee races, or breeds, as well as a number of stable hybrids that are kept commercially and by hobbyists. These breeds differ with respect to both their visual appearance and their behavior and adaptation to the environment. The main characteristics of what are generally considered the most important honeybee breeds are discussed briefly here.

Most bees belong to a breed of the Italian or Ligurian honeybee (*Apis mellifera ligustica*). They are relatively light in color, with light brown or golden and dark bands. Beekeepers generally considered these bees to be good foragers. Moreover, they build up large stores of honey. This bee breed starts collecting nectar and pollen early and finishes late in the day. Italian bees are regarded as good comb builders, but some consider the comb to be less attractive than those of the other breeds. For the beekeeper, Italian bees are considered calm and relatively easy to work with. With respect to their reproductive behavior, Italian bees tend to continue brood rearing after the nectar flow has ceased. The result is a tendency to rob other hives of honey and the need for more food to survive the winter.

In spring, Italian bees are considered relatively slow at building up their colony. They are also less likely to swarm than some of the other breeds. With respect to their resistance to pests and diseases, Italian bees have some resistance to European foulbrood but are considered rather susceptible to other diseases. They also have a strong cleaning behavior, which is important in removing pests from their hives. The golden Italian bees are best suited to warmer climates and seem to withstand drought conditions better than other subspecies.

The next most popular breed in the world is probably the Carniolan honeybee (*Apis mellifera carnica*). This breed originated in Central Europe (Slovenia). The color of the worker bees can be described best as grey with lighter hair on the abdomen. Carnolian bees are easy to handle. With respect to their foraging behavior, Carnolian bees also start to collect nectar and pollen early in the day. More so than Italian bees, this breed is known to forage in the rain and at lower temperatures. The Carnolian bees are relatively sensitive to the seasons; when the weather changes in the autumn, the queen bee stops laying, and when the temperatures increase in spring, their colonies quickly increase. As a result of their response to the seasons, Carnolian bees need less food during winter. The size of the colony, too, is influenced strongly by environmental factors, including the availability of nectar and pollen. Carnolian bees are known to swarm easily when their hive becomes too small to contain rapidly expanding brood. This breed is recognized for being less susceptible to brood diseases than other breeds. Their foraging and reproductive behavior indicates that Carnolian bees have adapted to colder climates and mountain regions.

Another breed worth mentioning is the Grey Mountain, or Caucasian, bee (*Apis mellifera caucasica*), from the foothills of the Central Caucasus mountain region. The Caucasian bee is similar in appearance to the Carnolian honey bee. Caucasian bees are generally regarded as having the most docile temperament and as the easiest to handle. Similar to the Carnolian bees, Caucasian bees build large and strong colonies, but less so than the Carniolan bees. These bees, too, are more strongly influenced by the change of seasons than are Italian bees, and they survive on less food during the winter. Similar to the Carniolan bees, Caucasian bees stop rearing brood in autumn. They also forage earlier and on days with lower temperatures than do Italian bees.

Caucasian worker bees are said to have a longer tongue than most races, which enables them to access a wider range of nectar sources. This may have occurred in their adaptation to harsh environments. These bees are known to produce a lot of propolis, which makes this breed difficult to manage in framed hives. Similar to the Carnolian bees, Caucasian bees are adapted to colder climates and mountainous regions and are able to cope with rapid changes in temperatures.

Prior to the introduction of the Italian and the Carniolan bees in Europe, different varieties of the European dark bee

(*Apis mellifera mellifera*) had developed in response to local conditions. These varieties have now become rare, as most have interbred with imported breeds. Some are listed as rare domesticated animal breeds. In an attempt to combine the characteristics of different subspecies, stable hybrids have been developed, the most well-known of which is the Buckfast hybrid, first produced by Brother Adam of the Buckfast Abbey. He crossed many races of bees in the hope of creating a superior, disease-resistant bee.[2]

2. The information on bee breeds presented in this addendum is from Bailey 1982 & de Bruyn 1997.

7

Healthy Bee Colonies

"*These scientists think they are saying something truly important. They announce their results using big words, and all the rest of the people believe everything they say. If his type of thing continues in the course of history, the time will come when everything will have to wither and die. For the world is dependent on your ability to actually do something and not only to talk about things and create new abstract terms. Words must mean what is really there. Long ago there used to be a kind of science that was directly connected to the actual experience of those things under examination. Today we have a science that knows nothing anymore of direct experience. All it does is think up fantastic new word combinations, and this is possible only because a new authority has been added to the old one.*"

—Rudolf Steiner (44)

The decline in the health of bee colonies was already a matter of concern among beekeepers in Steiner's audience on December 10, 1923. One of those beekeepers commented, "These diseases were not as virulent as they are today, at a time when a larger quantity of honey is taken from the bees and artificial methods are used for feeding and breeding bees."

Mr. Müller disagreed: "It is more likely that such diseases simply were not noted as carefully. Also, there were not as many hives with very small populations, so people did not pay as much attention to these things."

Mr. Erbsmehl contributed to the discussion by suggesting, "A possible cause might be the use of synthetic, chemical fertilizers. Flowers show the negative effects they can cause" (83).

The diseases being discussed were most likely foulbrood and nosema,[1] the rise of which coincided with the introduction of moveable frames and associated management practices such as artificial breeding and feeding (see Baker 1948; Thür 1946; Warré 1948).

Review of Bee Diseases

Since that discussion in 1923, the health of bee colonies has continued to decline, and bee colonies have been increasingly affected by a range of pest and diseases. At this point (2009), the most serious of these are the Varroa mite (Varroa destructor) and colony collapse disorder (CCD).[2] Australia appears to be the only country still free of some of these problems. Others, however, have taken a their toll, including nosema disease, American foulbrood, the small hive beetle (*Aethina tumida*), and wax moths (*Achroia grisella* and *Galleria mellonella*). Australia's proximity to Asia and the recent discovery there of the presence of Asian honeybees (*Apis cerana*), natural carriers of Varroa mites, suggests that the arrival of Varroa destructor in Australia is probably inevitable.

The decline in the health of bee colonies since the beginning of the twentieth century seems to follow a pattern, whereby as soon as treatments for one disease have been found, a new pest

---

1. European and American foulbrood is a bacterium that infests the mid-gut of an infected bee larva; nosema is a microscopic protozoan that affects a bee's stomach lining.

2. CCD is the abrupt disappearance of all bees from a bee colony. The term was first used during a drastic rise in disappearances of honey bee colonies in North America in late 2006.

and disease occurs with increased ferocity. Meanwhile, conventional scientific research is at a loss to explain or treat CCD.

Steiner was critical of the way conventional science studies nature by analyzing one factor in isolation from the totality of interrelationships. He argued that this approach cannot lead to a true understanding of events in nature. He investigates the health of bee colonies from a different perspective, taking a holistic approach. Steiner views animal diseases as evidence of imbalances among the physical, etheric, and astral dimensions of an organism, suggesting that any cure should begin by restoring balance. Based on this understanding, viruses and bacteria are not the problem; their presence is a symptom, and various causes may well result in the same symptoms.

With respect to issues such as CCD, it may not be a single factor that is responsible, but a number of factors that combine to create a negative effect greater than the sum of the individual causes. It is interesting to note that (albeit based on limited research and anecdotal evidence) the impact of pests and diseases is far less or absent in colonies that are managed in a more sustainable manner (Conrad 2007; Hauk 2008; Schacker 2008).

Bee colonies themselves have developed a number of strategies to minimize pest and disease problems as part of their development and adaptation to a particular environment. The first is their selection of a colony's nest site. Under natural conditions, a colony selects a nest site that is well off the ground— about 3 to 35 feet (1 to 5 meters), protected from rain and wind, and with a relatively small entrance. These strategies enable bees to minimize the risk of certain predators entering their nests and to maintain a stable internal environment.

Another defence is the colony's ability to deter intruders with venom. Depending on the size of the intruder, the effect of bee stings varies from mild discomfort to death. Bees have also been known to physically remove intruders that are difficult to sting, such as ants and hive beetles. Other examples of ways bees deal with intruders were discussed during Steiner's lectures on bees. Müller described how a hornet's nest inside a beehive had been subdued by the nest scent and rendered harmless. Steiner related how a mouse that had died inside the hive had been covered by propolis, which dehydrated its body and eliminated the effects of decomposition.

Bee colonies use propolis, nest scent, and the antimicrobial qualities of honey to reduce molds and other fungi and bacterial diseases. Bee colonies also minimize the effects of pests by adapting to coexistence with pests. For example, Asian honeybees are able to live with Varroa mites, which are a serious problem for European honeybees.

These examples show the importance of appreciating the bee colony's natural ability to deal with pests and diseases. With this in mind, an effective approach to hive health focuses on supporting the bees' ability to heal themselves. The ability of a bee colony to develop such strategies depends very much on its ability to live in harmony with its environment.

The Impact of Beekeeping Practices on Colony Health

As with people, there is a growing recognition that stress and lack of proper nutrition affect an animal's ability to fend off pests and diseases. Minimizing stress means that beekeepers must work with and support the bee colony's natural processes in both the spiritual and the physical dimensions. Conventional beekeeping practices such as the use of framed hives, frequent long-distance migration, artificial feeding, the use of antibiotics and chemical treatments, and environmental pollution all disrupt the natural processes of a bee colony and cause stress. For example, framed hives negatively affect the bee colony by interfering in the relationships between the colony and the wider environment. Under natural conditions, nest scent is released into the environment only when a nest is destroyed. The release of nest scent signals that a bee colony has been seriously damaged and needs to be cleaned up. The release of nest scent when a framed hive is opened sends the same signal and attracts any creatures with an interest in cleaning up a damaged bee colony, including bees from other colonies. This signal is even stronger when accompanied by damage to the comb or spilled honey.

Moreover, the current industry practice of regularly replacing the queen bee also places unnecessary stress on a bee colony. Killing the existing queen bee prior to transplanting an artificially inseminated queen bee into a colony seriously disrupts the intricate relationships within the colony. Long-distance travel may also affect the health of the newly mated queen bee.

Bee Nutrition and Bee Health

The importance of nutrition for the health of humans, animals, and plants has long been recognized by conventional science. In his lectures, Rudolf Steiner, too, emphasized the role of nutrition in the occurrence and treatment of pests and diseases in bees. He stated that inappropriate nutrition disturbs the balance of the bees' juices and overstimulates their astral bodies, which negatively affect colony health. He explained that the bee's nutritional process of converting nectar into honey is achieved through the relationship between the bee's gastric juices and blood. The gastric juice is white and acidic; the blood is a slightly redder liquid and alkaline. Steiner identified these as essential components of bees and their functioning.

Steiner stated that, to ensure the health of bees, it is necessary to maintain or restore a balanced relationship between the bees' gastric and blood juices. A lack of acidity in the gastric juice disturbs the ability to convert nectar into honey. For example, Steiner mentioned that, when bees are unable to obtain nectar in years when nectar flows are low and they are forced to feed mainly on the honey dew, the health of the bees' juices is affected. The bees lose their vitality and become more susceptible to diseases. According to Steiner, under natural conditions it is relatively easy to revitalize the blood juice if the appropriate external conditions, such as climate, light, and temperature, are present. However, as most beekeeping no longer considers the bees' nature, this is no longer effective. Instead, we need research to learn how to improve the quality of the bees' blood juice. Future beekeepers will need to check the health of their bees regularly and make sure that they are able to continuously correct any negative effects (86–87).

As a possible solution to counteract the negative effects of conventional beekeeping practices and to ensure healthy bees, Steiner suggested establishing a small greenhouse for plants that attract bees throughout the year, making them available to the bees out of season (88). He stressed, however, that this approach should first be the subject of a research project to examine its benefits.

In response to Müller's example of using American clover (presumably a reference to *Trifolium pratense,* a variety of red clover) for this purpose, Steiner cautioned against undertaking such a project without a proper understanding of the benefits of certain plants for bees and the negative effects of others. He mentioned that American clover doesn't improve the bee's blood juice over the longer term, but it works as a stimulant, similar to the effect of alcohol on humans, stimulating bees to expend a lot of energy for a short period of time (89).

Steiner suggested plants that occur naturally in the bees' environment. He reiterated that we should be clear that bees have adapted and are connected naturally to certain localities. This must be taken into consideration to achieve long-term benefits (89).[3] Steiner suggested that helping nature in this way should restore a colony's balance and control diseases.

------

3. As part of the European colonization during the last few centuries, the European honeybee accompanied the spread of European agriculture. During this period the honeybee has adjusted to these new environments and this should be taken into consideration in the selection of plant species to be used as part of Steiner's experiment.

An additional strategy to support bees in healing themselves is to revitalize plant life and thereby the quality of the bees' diet. During Steiner's lectures, Erbsmehl expressed his concern about the effects of artificial fertilizers on flowers and the quality of the nectar and pollen bees gather. Because bees receive all their nutrition from flowers, the quality of those plants is crucial to their overall health.

The use of chemical fertilizers was also regarded by farmers at the time as responsible for reducing the vitality of their seeds and animals. This led Steiner to develop a system of agriculture that is independent of chemical inputs, but works in harmony with nature and spiritual forces, which he outlined in his course on agriculture. The key objective of biodynamic agriculture (as these methods have come to be known) is to revitalize the soil as the basis for life on Earth.

Since Steiner's lectures, the use of chemicals in agriculture and horticulture has increased to such a level that virtually every aspect of the growing cycle is associated with one chemical input or another. The risks associated with chemical use in agriculture have been voiced continuously since the 1960s (see Rachel Carson's *Silent Spring*), and more recently it has been linked to the occurrence of CCD. Evidence suggests that bee colonies managed without the use of chemicals are less likely to succumb to CCD (Shacker 2008). Keeping bee colonies in areas where biodynamic farming practices are used or by applying biodynamic practices to the areas where bees are foraging will result in healthy soils and revitalized plant life. This in turn revitalizes bee colonies as they consume the nectar and pollen from plants grown in these locations.

During Steiner's lectures, he was asked specifically about foulbrood, the disease whereby young larvae die before they hatch. He initially thought that it had to do with a defective composition of the queen's uric acid, but said that further research was needed (102). Hauk (2008) also indicates a failure of the queen bee's metabolic processes as the cause for this disease, suggesting that the problem is related to the quality of the queen bee's nutrition. He described how he has been able to prevent foulbrood in his colonies by avoiding practices such as artificial breeding and feeding. Hauk also mentions the importance of keeping the brood warm in the early spring as part of a preventative strategy, but does not connect this to the use of framed hives. The occurrence of foulbrood has also been linked to the introduction of these hives (see Baker 1948; Thür 1946; Warré 1948). To minimize the immediate effects of foulbrood, Hauk (2008, 78–79) describes a successful method used by Thun (1986), based on stimulating the bees' cleaning behavior.

Feeding Bees

Bees derive all their nutrition from a plant's nectar and pollen, and they benefit spiritually from the cosmic forces stored in the plants. A bee colony stores honey and pollen, which enables it to survive when no nectar and pollen are available. This means that a bee colony can build its numbers early in the season and does not need to wait for sufficient nectar and pollen in nature. By doing so, the bees can revitalize plant life immediately after winter. It could be said that the honey and pollen stored by the bee colony contain a "reserve" of spiritual and physical energies, which the bees use to benefit plants before they become abundant in nature.

People probably realized early on in the history of beekeeping that without honey stores bees will not survive when nectar flows are at a low and during winter or periods of drought. Thus, they ensured that bees were left with enough stores to see them through those periods. However if the amount of honey stored by a colony proved to be insufficient for surviving the winter, beekeepers resorted to feeding their colonies. Initially this was done in times of prolonged winters or after poor nectar flows, and this was still the practice among beekeepers in Steiner's audience in 1923. However, because beekeeping was being dominated increasingly by profit gain, and because honey is more expensive than sugar, honey and pollen were soon replaced by sugar as bee food during winter dormancy. The substitution of sugar for honey and pollen enables beekeepers to take more honey and obtain greater economic return from their bee colonies. Today, feeding bee colonies has become the dominant practice and a permanent feature of conventional beekeeping. In addition to replacing natural food stores, feeding bees is also used to administer medication, to manipulate bees to build up their colonies, and as part of the artificial breeding process.

Generally, bees are fed a syrup consisting of about sixty percent sugar and forty percent water. Only refined sugar is used, as the impurities in natural and less refined sugar can result in dysentery. As an alternative to sugar syrup, bees can be fed sugar candy, a more crystallized and concentrated form of sugar syrup. If insufficient pollen is available because the beekeeper has taken it, the sugar syrup is supplemented with artificial pollen. Artificial pollen is generally made from soy flour, natural pollen, and brewers' yeast mixed with water and honey (de Bruyn 1997).

Review of Feeding Bees

During the lectures, Müller asked Steiner about his method of feeding bees artificially during times of emergency. Müller described how he prepared food by adding thyme, chamomile tea, a little salt, and about 11 pounds (5 kg.) of sugar to a gallon (4 liters) of water. He said that this method requires some preparation: "We have to clean out everything in the apiary, even the combs with unhatched bees, because the bees will otherwise get dysentery" (37).

Steiner was able to offer a bit of information on the value of this feeding practice. He observed that Müller's use of herbs was not very different from the basic principles of Steiner's own homeopathic medicines, but said that feeding sugar doesn't make much sense, as bees feed on nectar and pollen and are not used to eating sugar as part of their normal diet. This is why bees develop dysentery when fed sugar without an alternate form of nourishment. Steiner explained that bees (and other animals) try to transform the nourishment they are given into something they can use. Bees try to convert the sugar into a kind of honey, but this is less efficient than consuming natural honey, and only strong bees can do it (38).

Steiner also explained the value of the other ingredients in Müller's mixture. Plants contain a lot of starch, which tends to transform itself into sugar. Chamomile juice works on the starch and directs the sugary juice in the plant to become more like honey, and providing bees with chamomile tea supports this. It transforms the sugar and makes it easier to digest, since the bees do not have to perform this process in their own bodies. The substance in the chamomile plant exists in every plant with honey nectar, but chamomile contains more of it. Steiner indicated that this is why chamomile is an unsuitable source of honey (38). He added that the benefit of adding salt to the syrup is also beneficial, as salt spreads quickly through the body and helps make food more digestible (39).

Steiner emphasizes the importance of understanding the role of nutrition in the preparation of artificial bee food. Sugar, like honey, is sweet and of plant origin, and to assume therefore that it would be a viable substitute for honey might seem reasonable. However, bees do not naturally consume sugar, and when sugar is examined spiritually it becomes clear that it is not an appropriate substitute for honey.

The nectar from which bees produce honey is derived from the plant's flowers. Sugar, on the other, hand is derived from the plant's stem (cane sugar) or root (beet sugar). These are parts of the plant that bees do not use as part of their natural diet. At the spiritual level, too, bees have a connection to the flowers of the plant but not to the stem or root of the plant. Furthermore, stems and roots are formed very differently from the flowers and are created under the influence of different etheric formative forces. It is not surprising, therefore, that, from a spiritual perspective, sugar would cause difficulty with digestion. Instead of using cane or beet sugar it would be worthwhile to research the use of fructose in artificial bee food, as this is created under the influence of similar etheric formative forces are as the nectar and pollen.

When honey and sugar are examined materially, the chemical analysis of sugar, nectar, and honey shows that nectar and honey consist mainly of fructose and glucose, which are classified as monosaccharide and contain only small amounts of sucrose, a disaccharide and the main ingredient of sugar. The substitution of sugar for honey is an example of what Steiner referred to as anthropomorphizing nature, the application of a human-centric perspective. Just because both honey and sugar taste relatively similar to humans and can be used relatively interchangeably to sweeten our food, we should not assume that this applies to bees. Both spiritually and physically, honey is different from sugar.[4]

A Way Forward: Supplementary Bee Feeding

During winter, bees feed on the honey and pollen stored in their comb, a process that has developed as part of the bee colonies' adaptation to their environment. Steiner acknowledges that it may be necessary to artificially feed bees to prevent starvation, but if this is needed, bees should be fed their natural diet: honey and pollen.

Hauk (2008) discusses the implications of this for the timing of the honey harvest. He writes that, to ensure bees have access to honey and pollen throughout their period of

---

4. Human beings, too, have difficulty digesting sucrose, it needs to be broken down before it can be absorbed and used in the human body. Glucose on the other hand can be absorbed directly.

dormancy, honey should be harvested only after this period and when nectar and pollen can be found again in nature. Only during years with good nectar flows is it possible to extract honey during the summer. Under normal conditions, there should be no need to feed bees artificially, unless nectar flows have been low and the bees are at risk of starvation. If honey is supplied artificially, it is important that it is free of chemical pollutants and of good quality from a Demeter or certified organic source.

The International Demeter Standards for Beekeeping and Hive Products (2008) also emphasizes that honey and pollen are the natural food for bees, but allows the use of artificial feeding under a number circumstances—for example, to help bee colonies to survive the winter. The Standards state one should to aim use honey, but where this is not possible the food must contain at least five percent honey by weight from a Demeter-certified source. It also states that chamomile tea and salt should be added, which should be organic and preferably biodynamic.

The Standards also permit artificial feeding of emergency rations, for example prior to the first honey flow of the season. This must follow the same steps as artificial feeding during the winter. If emergency feeding is required later in the season and before the last harvest of the year, only Demeter honey should be used. The use of sugar is not allowed in such rations. Supplementary feeding may also be done to build the strength of swarming bees and those remaining behind, as long as the process is the same as that used for overwintering.

The artificial food set out in the Demeter standards seems to be based on the process that Müller used at the time of Steiner's lectures. This formula has also been adapted slightly by other beekeepers such as Gunther Hauk, whose bee food is based on a mixture of honey, sugar, and herb teas, including chamomile, sage, and a little salt, which meets Demeter requirements (Hauk 2008, 49).

Migration

Very early beekeepers discovered that it is possible to move a bee colony as long as it is protected within its nest cavity, sufficient worker bees accompany the colony, and enough stores are present to feed the colony while it adjusts to the new circumstances. This discovery enabled people to take advantage of different flowering periods and locations. The modes of transport of that time, however, meant that the distances were relatively small (see Crane 1999).

Until the late nineteenth century, it was not practical to transport bee colonies over long distances. With the widespread introduction of mechanized transportation, commercial migration of bee colonies started to be applied on a large scale. The benefit of migratory beekeeping for honey production was enormous; it enabled a country like Australia to double its honey production every five years during the 1940s (Crane 1999). Unfortunately, the practice of moving a large number of bee colonies over long distances coincided with an increase in pests and diseases. This may have happened because the movement of bees over large distances enabled pests and diseases to spread more easily at a time when the bee colony's vulnerability was increased by the stress of long-distance travel (Conrad 2007; Schacker 2008).

Review of Migratory Beekeeping Practices

Under natural conditions, the only time a bee colony moves locations is during its reproductive cycle, when it is swarming. Once a bee colony has found a nest cavity, it rarely leaves it voluntarily.[5] Individual bees may travel substantial distances in search of nectar and pollen, but bee colonies in their natural state do not migrate. Bees by their very nature are not migratory animals.

---

5. Bee colonies may decide to swarm if their nest is invaded by a pest and they perceive that they cannot fight this pest.

At the time of Steiner's lectures, beekeepers in some European countries moved their hives to take advantage of the different flowering periods. The large-scale movement of hives over hundreds or thousands of miles, which is characteristic of modern beekeeping, did not occur, and thus Steiner did not comment on it. He did observe, however, that bee colonies become used to, and are very strongly attached to, the region to which they belong, as shown by variations in the morphology of bees from different areas (89). The process used to move beehives long distances is unnatural and frequently requires bees to adapt to new climatic and other conditions. When examined spiritually, the inability of a bee colony to adapt to a particular environment because it has been moved regularly can undermining its long term well-being.

Steiner also warned about the bees' diet. A beekeeper should not feed bees on flowers and pollen that are foreign to them. Bees become used to the flora of particular areas, and feeding bees nectar derived from areas foreign to them requires extra energy for the bees to convert and digest unfamiliar nectar and pollen and creates instability in their digestive system. Steiner stated that nectar and pollen from foreign sources will not be beneficial in the long term (90). Steiner's comments were reinforced by Müller, who said that European bees die when fed American honey.

The current International Demeter Standards (2008) allows seasonal hive movements. However, it does not specify or make recommendations with respect to the frequency or distance of this throughout the season.

Little is known about the impact of long distance migration on bee colonies. Perhaps there are limits, both in distance and in frequency of moves, within which bee colonies can be moved without too much harm.[6]

---

6. Ironically, perhaps the artificial feeding of migratory bees minimizes their need to adjust to new food sources every time they are moved to a different environment.

A Way Forward: The Migratory Beekeeper

Many crops, especially in the U.S., depend on migratory beekeeping for pollination, so any sudden abandonment of the practice seems impractical at first glance, as it would result in shortages of some food crops. An alternative to migratory beekeeping could be for the horticulturists and farmers who rely on migratory beekeeping for pollination to keep their own bee colonies. Farmers and horticulturists who are not interested in becoming beekeepers could engage the services of a beekeeper to manage the colonies. To make such a proposition viable economically, the beekeeper could be the one who travels from apiary to apiary to care for them.

For this alternative to be practical, enough nectar and pollen sources need to be available throughout the year for bees to feed on after the pollination of the main crops has finished. This would require horticulturists to diversify their crops and other vegetation on or around their land. It could even create opportunities for new income by growing different crops throughout the seasons while creating greater biodiversity. Unfortunately, the high levels of pesticide and other chemicals used in conventional horticulture form a serious barrier to this alternative to migratory beekeeping on most conventional farms.

Other Possibilities for Revitalizing Bee Colonies

In lecture 6 of Steiner's *Agriculture: Spiritual Foundations for the Renewal of Agriculture*, he suggested a number of strategies to manage pests. For example, they can be managed by developing an understanding of alternative environments to which the pests are attracted and then creating such an environment in the vicinity of the bee colonies.

Steiner also discussed how to use cosmic forces and the zodiac to affect insect life. He focused mainly on the preparation of "insect peppers," remedies for managing insect pests. The application of the peppers can be used to reduce the presence of the insect pests that affect bee colonies, such as the small hive beetle and possibly even the Varroa mite.

The research and reasoning behind the process suggest that it should also be possible to develop remedies or preparations to revitalize bee colonies and thereby enable them to stave off pests and diseases. However, to date this possibility has not been explored, and there is little evidence of experimentation in this area.[7] Further information on the preparation of so-called insect peppers can be found in Steiner's *Agriculture*.

It is important to reiterate here that the incidence of pests and diseases are the result of disturbances in the natural balance of a bee colony and its environment, and that the remedies discussed here will not have any lasting effect if they are not accompanied by beekeeping practices that restore balance. Referring back to the Varroa mites, they are considered pests only in the context of the European honeybee; the Asian honeybee is able to coexist with the mites without any negative effects.

Discussion

In natural conditions, bee colonies develop strategies to minimize the impact of pests and diseases—for example, the unique way a colony constructs combs to create and maintain a stable internal environment, the selection of nest sites, and the use of venom to minimize the negative impact of external disruptions. The introduction of framed hives and management practices such as artificial breeding interfere with the natural processes of bee colonies and negate the effectiveness of their strategies. Such practices have been associated with an increase in pests and diseases.

Today the life of a conventional bee colony is characterized by long-distance travel, manufactured food, and life in

---

7. In his agriculture lectures, Steiner stated that the fertility of insects is governed by the forces of the Sun as it moves through the zodiac. He also indicated that the element of fire can be used to destroy fertility, and that the element of water is closely associated with enhancing fertility and can be used in the development of biodynamic preparations.

prefabricated, unnaturally shaped dwellings. The life of a bee hive is more like that of a busy executive than one in keeping with its nature. This fact makes it difficult for bee colonies to maintain and use their naturally developed mechanisms for fending off pest and diseases.

Conventional science does not generally link the occurrence of pests and diseases to human interference with a bee colony's natural processes. The conventional approach to minimizing and eliminating the impact of pests and diseases is based on linear thinking with a narrow focus on cause and effect. This approach has been unsuccessful in explaining and offering solutions to the most recent ailments affecting bee colonies, including colony collapse disorder.

Spiritual science examines pests and diseases holistically, which Steiner found to be the result of imbalance in the bee colony's astral forces. To address this problem, it is essential to restore the colony's natural processes and develop strategies to compensate for the negative effects of conventional management practices.

With respect to restoring the natural balance of bee colonies, Steiner emphasized the role of nutrition and the importance of adopting the perspective of the bee instead of anthropomorphizing bees and substituting sugar for honey simply because both taste sweet.

In restoring the natural balance of bee colonies, it is also important to examine the effects of conventional beekeeping practices such as migration, as these are likely to increase a colony's stress levels by forcing it to derive nutrition from plants in unfamiliar locations. Steiner found that such practices negatively affect a bee colony's well-being. He suggested a number of strategies to restore the health of bee colonies, including increasing their access to beneficial plants and enabling them to forage in locations where biodynamic preparations are used.

To address the immediate effects of pests and diseases, it may be necessary to employ interim measures. In doing so it is important to realize that these by themselves do not provide a cure and do not address the causes of the disease or pest outbreaks.

Steiner also suggested that the effects of pests can be minimized by providing environments that attract them away from the hive. In addition, his development of methods to affect insect fertility could be explored both by developing pepper applications that minimize the presence of pests and by preparations that improve bee health.

In each case, it is important to understand the well-being of bee colonies holistically and to assess the impact of any intervention on the totality of relationships within the bee colony and its relationships with the wider environment.

ADDENDUM BEEKEEPING PRACTICES AND HONEY

Previous chapters discussed the important qualities that Steiner attributed to honey, which make it so beneficial for human consumption. These qualities arise from the way that bees transform plant nectar into honey, a process that infuses honey with astral qualities. The beneficial qualities of honey also result from the absorption of cosmic energies while stored in the hexagonal honeycomb cells built by the bee colony.

In light of the important role of bees and the honeycomb in the formation of honey, it is likely that the conventional beekeeping practices discussed previously, which disrupt the natural processes of the bee colony, also have negative effects on the qualities of the honey produced by such bee colonies. Little research has been undertaken to date to explore this relationship. The table opposite shows the differences between conventional beekeeping and beekeeping practices based on spiritual science as they relate to the formation and harvest of honey.

HONEY QUALITY: CONVENTIONAL BEEKEEPING VS. BEEKEEPING PRACTICES BASED ON SPIRITUAL SCIENCE

CONVENTIONAL BEEKEEPING	BEEKEEPING BASED ON SPIRITUAL SCIENCE
The beekeeper pays little attention to the pollution levels or toxicity of the environment in which bees gather nectar.	The beekeeper considers the toxicity and pollution of the environment in which bees collect nectar.
Bees construct honeycombs in frames and hives made of material treated with chemical paints and other synthetic substances.	Bees construct honeycombs in hives made from materials that are treated with only natural products.
Honey is stored and ripened in honeycombs that bees have constructed on recycled wax or plastic foundations.	Honey is stored and ripened in honeycombs that bees have constructed naturally.
Honey is collected whenever the supers are full.	Honey is stored and ripened in natural honeycombs for 3 and 9 months.
Honey comes from bees that are fed artificially.	Honey comes from bees that are fed their natural diet.
Honey comes from bees that may have been treated with antibiotics and other chemicals to minimize the effects of pests and diseases.	Honey comes from bees treated with only organic and biodynamic preparations to maintain their health.
Honey is heated to facilitate extraction from the comb.	Honey is extracted at low temperatures.

8

THE ROLE OF THE BEEKEEPER

"The feeling that they [the beekeeper and the bee colony]
belong to each other is based on the fact that a great
wisdom lives in the beehive. It is not simply an assembly
of individual bees; the beehive really has its own very
specific soul."

—RUDOLF STEINER (185)

The previous chapters examined a series of lectures by
Rudolph Steiner on the nature of bees in 1923. The purpose
has been to determine how his research can be used to improve
the practice of beekeeping. The application of spiritual science
to understanding bees and beekeeping brings out a number of
aspects that are generally ignored by conventional beekeepers. The most prominent of these are the role of economic and
political conditions in determining beekeeping practices; the
need to identify and study the wisdom and spiritual element
as expressed physically in nature; the importance of studying
nature holistically within an evolutionary context and with a
focus on relationships that are cooperative and mutually beneficial; the role of cosmic influences; the role of qualitative factors
in understanding nature; and awareness of the risks of anthropomorphizing when interpreting nature.

Referring to the impact of economic and political factors, Steiner showed that beekeeping in Europe was beginning to change radically because of the increasing commercialization of society. Those changes were driven by the need to improve honey production to increase financial returns, leading to the adoption of beekeeping practices in support of short-term economic goals such as artificial bee breeding. Discussions between beekeepers at that time, however, indicate that such practices were not welcomed universally. Some beekeepers feared they would undermine the well-being of bees in the longer term. Against this background, Steiner insisted that beekeepers should be aware of current economic and social dimensions and how they shape the development of beekeeping practices.

Steiner pointed out that the way in which the effects of these practices are perceived really depends on how they are studied. He stated that conventional science does not acknowledge the existence or relevance of the spiritual element in shaping the material world; it has a tendency to zoom in while seeking to understand nature by reducing it to the smallest components, which are studied in isolation from their context. Steiner stated that, consequently, conventional science provides only a partial understanding of nature.

By contrast, Steiner discussed a holistic approach to studying nature, which acknowledges that the natural world cannot be fully understood without studying its spiritual basis. Spiritual science examines nature from the perspective of the whole organism and avoids anthropomorphizing natural phenomena. In his lectures on bees, Steiner showed how a spiritual approach emphasizes the importance of cosmic influences, the shape of comb cells, and variations in the duration of bees development for a true understanding of bees and beekeeping.

Steiner's research findings identified a number of principles that are important for understanding the nature of bees and the effects of beekeeping practices. Central to this understanding is the notion that a bee colony is a closely integrated entity with its own consciousness, being more like an organism than a collection of individual bees. Any interference should be avoided, as it disturbs the organism internal relationships and processes. He also stated that the relationship between a bee colony and the wider environment consists of an integrated system built on cooperation and mutual benefit rather than on competition. Bees benefit from plants, which provide them with all their nutrition and the cosmic forces needed for their continued survival. In turn, bees benefit plants by playing a vital role in ensuring that plant life on Earth continues by revitalizing and stimulating plants and enabling their long-term reproduction, both spiritually, by bringing cosmic warmth to the seed during pollination, which enables it to ripen, and physically, exchanging bee venom with nectar.

The benefits of this relationship between bees and plants to humans is clear. Humans benefit from the continued vitality of plant life on Earth and the production of fruit and other food crops. In addition, we benefit from the consumption of honey, which not only provides an enjoyable taste experience, but also by provides solidity to our bodies, enabling us to partake in the cosmic forces stored within it that work on our astral bodies (as is presented in the diagram). Thus, while human beings benefit from keeping bees, the relationship seems to be rather one-sided. This raises the question: What do we provide in return?

To ensure that bees will be able to fulfill their important role in nature in the future requires human beings to focus not on short-term economic gains, but on promoting the long-term survival of bees and making this the real objective of beekeeping practices. Beekeepers must take responsibility for the well-being of bee colonies by changing the relationship between bees and human beings—from one of exploitation for short-term economic gain to one of mutual benefit.

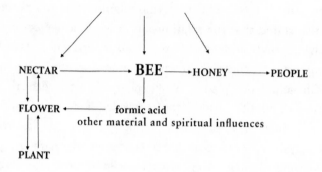

In addition to the ethical dimension of this responsibility, there are very practical reasons for doing so. Failure to ensure the well-being of bees will greatly affect plant life and food production and, thereby, the future of humanity.

However, the impact of economic and social conditions on beekeeping since the beginning of the twentieth century has led to a situation that Thür described as early as 1946, in which most beekeepers no longer understand the natural requirements of bee colonies and are thus unable to assess the benefits and harmful effects of modern beekeeping practices. Conventional science continues its focus on developing practices that maximize short-term economic benefits through techniques that manipulate the colony's natural processes, including bee breeding and artificial feeding. Such innovations are generally promoted by government agencies and advocates for industry.

Beekeepers can work toward the well-being of bees by basing their practices on Steiner's work. This requires an understanding of how the wisdom of bees has developed over time, both spiritually and physically. By understanding Steiner's research, it is possible to develop a different set of beekeeping techniques based on observation and diagnostic skills that are very different from conventional approaches.

At present, little is known about the impact of hive shapes and materials on the well-being of bee colonies, or the influences of cosmic forces on bee behaviors such as swarming. Similarly, the effect of biodynamic practices on the quality of nectar and pollen, the bees' natural food, and on the quality of honey for human consumption is not well understood. With respect to the management of bee colony health, little is known about the possible use of biodynamic preparations for minimizing the effects of pests and diseases and for revitalizing bee colonies. Moreover, the impact of conventional beekeeping practices such as migration on the long-term well-being of bees has not been well researched. Little is known about the impact of unnatural influences such as electromagnetic waves, nanotechnology, and genetically modified organism on bee colonies, and by extension, how to compensate for such effects through one's beekeeping practices.

The table on the following pages summarizes some key differences between conventional and spiritual approaches to beekeeping. Each of the categories poses issues and research questions that extend beyond the scope of this book and its objectives.

In addition to researching, developing, and implementing holistic beekeeping practices, counteracting established views and conventional beekeeping practices is not an easy task. This could necessitate pressing for changes in government legislation to enable alternatives to the beekeeping practices sanctioned by conventional science. The decline of the world bee populations and the seeming inability of conventional approaches to address issues such as CCD will, one hopes, help create greater acceptance of approaches to beekeeping that restore and maintain the integrity of bee colonies.

The development of alternative beekeeping practices highlights the challenge of meeting the needs of both the bee colony and the beekeeper. Conventional beekeeping serves primarily the economic objectives of our time; perhaps it is time for the pendulum to swing the other way.

Comparison of Key Conventional and Spiritual Science Beekeeping Practices

Practice	Conventional beekeeping	Spiritual science
Hive design	Square or oblong boxes; moveable frames; wax foundations; empty super placed on top of brood chamber. Little regard for maintaining brood temperature and nest scent.	Closed fixed frame; natural foundation; observation window if anything; old brood chambers become honey stores; new brood chamber placed below existing brood chamber. Hive shape as cylindrical as possible. Size 1/3 to 3 1/2 cu. ft. (10–100 liters), ideally about 1 1/2 cu. ft. (40 liters).
Reproduction	Queens: breeding and rearing.	Swarming: induce swarming of better hives. Examine the effect of cosmic influences such as those of the Moon and Venus on swarming.
Feeding	Artificial.	Hives own honey stores; in emergency use nutrition based on homeopathic honey treatments.
Honey harvest	Impact of disruption on colony receives little consideration.	Minimize disruption.

PRACTICE	CONVENTIONAL BEEKEEPING	SPIRITUAL SCIENCE
HONEY	The quality of honey is examined chemically and based on its culinary value.	The quality of honey is based on the effects of cosmic influences.
MIGRATION	Travels.	Limited or no travel; hives ideally remain stationary.
DISEASE TREATMENT	Antibiotics and chemicals.	Pests and diseases are a disturbance of balance, recreate balance, no chemical pesticides use of homeopathic treatments and biodynamic preparations.
FIELDS	Orchards and fields are characterized by monoculture and the use of chemical fertilizers, pesticides, and herbicides.	Diverse range of plants available throughout the season; grown biodynamically.

Fortunately, initiatives exist for the development of alternative beekeeping practices that are more in harmony with the wisdom of the bees. For example, people such as David Chandler and Ross Conrad are actively researching and promoting the use of top-bar hives and organic beekeeping. Gunther Hauk and David Weiler have worked for years in the biodynamic tradition. David Heaf is playing an important role in making information on alternative beekeeping practices such as those developed by Warré and the observations of Thür available to the English-speaking world. Not all of those who work with and promote alternative practices are guided by a spiritual-scientific perspective, but they nonetheless give priority to the well-being of the bee colonies in their beekeeping practices.

In all of this, it is important to bear in mind that, if we wish to manipulate the natural processes of the honeybee for our own gain, we must still maintain the integrity of the natural system, both physically and spiritually. This requires a holistic understanding of the bee colony's wisdom and a willingness to work with the natural processes of the bee colony.

This issue is very urgent at present, and in light of the importance of bees for maintaining and revitalizing plant life, for the Earth, and for people, the implementation of enlightened beekeeping practices should override the importance of increased economic gain. The present decline in world honeybee populations adds weight to Steiner's observation that humanity has a choice: Relearn the whole relationship between nature and the cosmos or to let nature die and degenerate. This choice is even more urgent today than it was at the time he made this point during the early twentieth century.

Appendix: Hive Design Features

The following features are presented to assist in the design and development of beehives that respect the integrity of the bee colony. The goal of these hive design features is to work in harmony with the wisdom that underlies the bee colonies activities. These features are based on the research findings of Steiner and others with a similar sensitivity to the nature of bee colonies and the Demeter Standards for Beekeeping and Hive Products.

- Hives constructed from materials that enable the colony to remain dry, and that would be available as a natural choice for bees to nest in. Little information is available on the impact or preference bees have for horizontal or vertical hives, or on whether this preference differs depending on type of bee, the climate, or other local conditions.
- Avoid metals, plastics and other synthetic and chemical materials such as paints and sprays in hives.
- Minimize interference with the bee colony throughout the year; any interference should ensure that temperature levels and nest scent are maintained.
- Allow natural comb construction in both shape and cell size to allow for uninterrupted development of the brood and freedom of movement of all the bees throughout the hive.

- Develop hive dimensions in keeping with climatic conditions; a hive volume of about 1 1/2 cubic feet (45 liters) and a hive entrance of about 1 1/2 inches (35 millimeters) can be taken as starting points; additional investigation is needed for further information on this matter.

- Position the hives in keeping with the bees' natural nesting instincts; conventional hives are usually placed relatively close to the ground, whereas a colony's natural preference is to build their nest away from predators, at least 3 feet (1 meter) above the ground. Hives should be located at least above low-growing plants to allow air circulation around the hive entrance and allow bees to fly freely in and out of the hive.

- Insure that the hive is stable to prevent the entry of pests.

- Bees require access to clean, unpolluted air, water, and flora.

- Position the hive with respect to the Sun and planetary influences and the Earth's magnetic fields. Additional research is needed on the effects of planetary influences on the colony, particularly those of the Sun and Venus; as Steiner pointed out, bees are Sun animals and are influenced strongly by Venus.

BIBLIOGRAPHY

Bailey, Fred, 1982. *Beekeeping in Australia: For Pleasure and Profit*, Richmond, Vic.: William Heinemann Australia.

Baker, C. T. G., 1948. *Understanding the Honeybee*, London: Biodynamic Agricultural Association, Rudolf Steiner House.

Carson, Rachel, 1962, *Silent Spring*, Houghton Mifflin, Boston, MA.

Crane, Eva, 1983. *The Archaeology of Beekeeping*, Ithaca, NY: Cornell University Press.

————, 1999, *The World History of Beekeeping and Honey Hunting*, Routledge, New York.

Conrad, Ross, 2007. *Natural Beekeeping: Organic Approaches to Modern Apiculture*, White River Junction, VT: Chelsea Green Publishing.

de Bruyn, Clive, 1997. *Practical Beekeeping*, Ramsbury, UK: Crowood Press.

Demeter International e.V, 2008. "Standards for Beekeeping and Hive Products," Germany.

Hauk, Gunther, 2008. *Toward Saving the Honeybee*, Junction City, OR: Biodynamic farming and Gardening Association.

Keats, B., 2009. *Antipodean Astro Calendar*, published by Brian Keats.

Laidlaw, Harry. H. & Robert E. Page Jr. 1997. *Queen Rearing and Bee Breeding*, Kalamazoo, MI: Wicwas Press.

Morse, Roger, A. & Ted Hooper, T, 1985. *The Illustrated Encyclopedia of Beekeeping*, New York: Dutton.

Schacker, Michael, 2008. *A Spring Without Bees: How Colony Collapse Disorder Has Endangered Our Food Supply*, Guilford, CT: Lyons Press.

Steiner, Rudolf. 1993. *Agriculture: Spiritual Foundations for the Renewal of Agriculture*, Biodynamic Farming and

Gardening Association, Kimberton, PA (also available as *Agriculture Course: The Birth of the Biodynamic Method*, London: Rudolf Steiner Press, 2005).

——, 1994. *How to Know Higher Worlds: A Modern Path of Initiation*, Hudson, NY: Anthroposophic Press.

——, 1995. *Nature Spirits: Selected Lectures*, London: Rudolf Steiner Press.

——, 1998. *Bees*, Great Barrington, MA: Anthroposophic Press.

——, 2008. *Goethe's Theory of Knowledge: An Outline of the Epistemology of His Worldview*, Great Barrington, MA: SteinerBooks.

Thun, Matthias, 1986. *Die Biene: Haltung und Pflege*, Biedenkopf, Germany: Aussaattage M Thun-Verlag.

Thür, Johann, 1946. "Beekeeping: Natural, Simple and Ecological," David Heaf, trans.; available for download at http://www.biobees.com/downloads/thur.pdf.

Warré, Abbé Émile, 1948. *Beekeeping for All* (trans. by Pat Cheney and David Heaf from *L'apiculture pour tous*, 12th ed.), self-published, 2007; available as a free download at http://thebeespace.files.wordpress.com/2008/12/beekeeping_for_all.pdf.

Weiler, Michael, 2005. "Biodynamic Beekeeping: Questions Put to Biodynamic Beekeeping Consultant Michael Weiler," http://www.biobees.com/downloads/BD_FAQ_Weiler.pdf.

——, 2006. *Bees and Honey: From Flower to Jar*, Edinburgh: Floris Books.

Resources for Beekeeping

This short list of resources is intended to help those unfamiliar with the world of holistic beekeeping. An online search will turn up many more resources. One of the best resources for anyone just starting out is to connect with local beekeepers to find out the kinds of conditions and needs that are particular to your area. There are also a number of online blogs, where beekeepers share their passion and problems. A good deal of information can be found, for example, at the Natural Beekeeping Network: www.biobees.com/forum.

Many of the web sites listed here provide a good deal of general information and further links that will be of value regardless of where you live and keep bees or otherwise engage in aspects of biodynamic gardening and agriculture.[1]

Australia

Biodynamic Agriculture Australia
    PO Box 54, Bellingen, NSW, 2454, Australia
    Tel: +61 02 6655 0566
    Email: bdoffice@biodynamics.net.au
    www.biodynamics.net.au
The Bio-Dynamic Agricultural Association of Australia
    C/- Post Office, Powelltown, Victoria, 3797, Australia

---

1 This list of resources is for the reader's further research and does not necessarily imply that the author or SteinerBooks endorses the methods and philosophies of the organizations listed.

Tel: +61 03 5966 7333
www.demeter.org.au

Bio-Dynamic Research Institute: Demeter certification
    C/- Post Office, Powelltown Victoria 3797 Australia
    Tel: +61 03 5966 7333
    www.demeter.org.au

The National Association for Sustainable Agriculture,
    Australia (NASAA)
    PO Box 768, Stirling SA 5152
    Unit 7/3 Mount Barker Road, Stirling SA 5152
    Tel: +61 8 8370 8455
    Email: enquiries@nasaa.com.au
    www.nasaa.com.au

Biodynamic Education Centre
    PO Box 1017, Queanbeyan, NSW, 2620, Australia
    Tel: + 61 2 6297 2729
    www.biodynamiceducationcentre.com

CANADA

Association de Biodynamie du Québec
    375, rang des Chutes, Ham-Nord, Québec J4Y 2R2
    www.biodynamie.qc.ca

Bio-Dynamic Agricultural Society of British Columbia
    1764 Hwy 3, Cawston, BC V0X 1C2
    Tel: (250) 499-2596
    Email: bdcertification@yahoo.ca

Certified Organic Associations of British Columbia
    202 3002 32nd Ave, Vernon BC V1T 2L7
    Tel: (250) 260-4429
    Email: office@certifiedorganic.bc.ca

Society for Bio-Dynamic Farming & Gardening in Ontario
    RR #2, Elora, Ontario N0B 1S0
    Tel: (519) 843-6822
    Email: mailbox@biodynamics.on.ca
    www.biodynamics.on.ca

New Zealand

Bio-Dynamic Farming and Gardening Association
PO Box 39045, Wellington, New Zealand
Tel: +64-4-589 5366
Email: info@biodynamic.org.nz
www.biodynamic.org.nz

Organics Aotearoa New Zealand (OANZ)
Petherick Tower, 38 Waring Taylor Street
PO Box 1926, Wellington, New Zealand
Email: info@oanz.org.nz
www.oanz.org.nz

Taruna College
PO Box 8103, Havelock North, New Zealand
Tel: +64 6 8777 174
Email: info@taruna.ac.nz
www.taruna.ac.nz

United Kingdom & Europe

The Biodynamic Agricultural Association
Painswick Inn Project, Gloucester Street,
Stroud, Glos GL5 1QG
Tel: 01453-759501
Email: office@biodynamic.org.uk

Federation of Irish Beekeepers' Associations (represents 45 local member associations in Ireland); much general information is available on their web site: www.irishbeekeeping.ie.

United States

The Biodynamic Farming and Gardening Association
25844 Butler Road, Junction City, OR 97448
Tel: (541) 998-0105
www.biodynamics.com (Links on this site list regional BD associations, training, and CSAs.)

Demeter USA
PO Box 1390, Philomath, OR 97370
Tel: (541) 929-7148
www.demeter-usa.org

The Melissa Garden—Holistic Beekeeping and Honeybee
     Sanctuary in Healdsburg, CA
     www.themelissagarden.com.
Partners for Sustainable Pollination
     1828 Beaver Street, Santa Rosa, CA 95404
     www.pfspbees.org
Rudolf Steiner College—Beekeeping Workshops
     9200 Fair Oaks Boulevard, Fair Oaks, CA 95628
     Tel: (916) 961-8727;
     steinercollege.yellowpipe.com/?q=node/163.
Spikenard Farm— Holistic Beekeeping, Training & Education
     RR3, Box 167; Carrollton, Il. 62016;
     Tel: (217) 942-3732
     www.spikenardfarm.org.

WORLDWIDE

Demeter-International: www.demeter.net; Demeter International
     has sixteen member organizations from Europe, America,
     Africa and New Zealand. Demeter-International represents
     more than 4,200 Demeter producers in forty-three countries.
     Visit www.demeter.net for information in a specific country.